TGIF Math

TGIF Math

Jim Elander

To order additional copies of this book, contact:
Xlibris Corporation
1-888-795-4274
www.Xlibris.com
Orders@Xlibris.com
131416

TABLE OF CONTENTS

A. Rodin's, *THE THINKER*, Golden State Park, San Francisco, CA

Why For The Teacher? 9
Prologue 13

Chapter 1. Statements are true or false, Conclusions
    are Valid or Invalid? 15
    Activity 1.   Predictions may be Wrong! 16
    Activity 2.   One Sided Paper or the Mobius Belt 17
    Activity 3.   Taxicab Geometry 17
    Activity 4.   Polygon Angle Sum 19
    Activity 5.   Curves from Straight Lines 19
    Activity 6.   What do you see? Illusions.....................21
    Activity 7.   More Illusions 22
    Activity 8.   Teen or Sr. Citizen 24
    Activity 9.   What do you see in the following pictures? 25
    Activity 10. How many times can a piece of paper be
                  folded in half? 25
    Activity 11. Looking for the easy way 25
    Activity 12. Why cars and boats tip over! 26
    Activity 13. All circles have the same measure for the
                  circumference. Your kidding! 27
    Activity 14. FedEx Logo 28

Chapter 2. Number Theory—Algebra—Reasoning
        Activities Potpourri 29
    Activity 1:   Which areas of study are important
                    for the future?...30
    Activity 2:   This problem will tell the student
                    if their answer is correct. 31
    Activity 3:   How to become a millionaire in 30 days. 31
    Activity 4.   How large is the college class? 32
    Activity 5.   How many of each were purchased?....................32
    Activity 6.   What number?...33
    Activity 7.   Dad, please send money!.................................33
    Activity 8.   Clock Arithmetic...33
    Activity 9.   1=0?? How can this be?..................................34
    Activity 10. Average speed (SAT or ACT question)................34
    Activity 11. Summer pay 35
    Activity 12. Diagonal of a 4th dimensional cube from a
                    mathematical viewpoint..............................35
    Activity 13. How many lanes is the coach planning for?.............36
    Activity 14. Saturday's pay 36

Chapter 3. Inductive Reasoning conclusions and related Problems........37
    Activity 1.   Preamble..39
    Activity 2.   Inductive reasoning 39
    Activity 3.   Points and Lines 40
    Activity 4.   Taxicab Geometry 41
    Activity 5.   N-gon and diagonals 42
    Activity 6.   The Tower of Hanior......................................43
    Activity 7.   10 Coin Problem 43
    Activity 8.   Cook's problem:...44
    Activity 9.   Given 4 fours (4,4,4,4) try to write the first
                    20 counting numbers using any operations or
                    combination of symbols.................................45
    Activity 10. All of your students have played checkers
                    (an assumption) so here is the problem. 45
    Activity 11. The game of Sprouts.......................................46
    Activity 12. What is wrong with this conclusion?....................47
    Activity 13. Interesting quotients 47
    Activity 14. The Way Out. 48

Chapter 4. Thinking and Reasoning Activities 50
    Activity 1.   The Declaration of Independence 51
    Activity 2:   The President's Advisors 52
    Activity 3.   The Cell Phone Problem? 53
    Activity 4.   Galileo and Gravity 54
    Activity 5.   A new type of number 54
    Activity 6.   Questionable Request and Ads..........................55
    Activity 7.   Implications and forms of Implications.............56
    Activity 8.   Ads and implications 57
    Activity 9.   The sum of the first N counting numbers.............59
    Activity 10. John's Walk 59
    Activity 11. Stack Problem...60
    Activity 12. One-sided paper or the Mobius Belt 60
    Activity 13. Who was guilty? (Indirect reasoning concept
                    is very important.) 61
    Activity 14. Text Message 62
    Activity 15. Magic Square...62
    Activity 16. Who did it? 62
    Activity 17. Clock arithmetic..63
    Activity 18. The eye deceives you...64
    Activity 19. Proof that 2 equals 1 or does it? 65
    Activity 20. Trip problem 65
    Activity 21. Activity for the day before a Federal Holiday............66
    Activity 22. How many triangles? 66
    Activity 23. Early out...67
    Activity 24. Highest score 67
    Activity 25. Summer jobs..68
    Activity 26. A weight problem, but a good buy!.......................68
    Activity 27. Ad Interpretation...69
    Activity 28. $1 - 0$ and or $\infty - 0$???? 71
    Activity 29. Easy Test...71
    Activity 30. The Binary System 71
    Activity 31. House and Dogs 72
    Activity 32. The Sad Romance of Miss Gon 72
    Activity 33. Interesting conclusions 74
    Activity 34. Sums, the easy way 75
    Activity 35. A Valentines Day Activity..................................75

Chapter 5. Additional Selected Activities for TGIF MATH Review 77

Bibliography...89

Quotes 95

* Ronan's *The Thinker* at Golden State Park in San Francisco, CA

WHY FOR THE TEACHER?

A set of Activities designed to improve your students SAT, ACT and other college or vocational school entrance exam scores, and more importantly your Decision Making Skills. Research has indicated that EVERYDAY CRITICAL THINKING SKILLS has scored the lowest on these tests. This program reviews and explains these skills and also incorporates some basic and new understandings relative to Decision Making Skills for everyday situations. Students' scores will go up!

The following statement was over the entryway to Plato's Academy.

LET NO MAN IGNORANT OF GEOMETRY ENTER HERE

Plato's statement reflects how important he felt Geometry was to everyday Decision Making and indicates why schools want students to have Geometry. (Geometry taught correctly will improve Decision Making skills. Plato knew this and Fawcett justified it in the 1930s (13th yearbook of the NCTM, NATURE OF PROOF). It is also suggested you read chapter 14 in Polya's MATHEMATICAL DISCOVERY-Vol.2 as to a method. This set of activities which you (the teacher) will carefully select from to fit the needs of your students, will prove beneficial and informative.

Rodin's THE THINKER (Golden Gate Park, San Francisco, CA)

Logical decisions are based on, undefined terms, defined terms, postulates, inverses, converses, contrapositives, using direct and indirect justifications, plus understanding their use and misuse. Keep in mind statements are truth or false and conclusions are valid or invalid.

These Activities are the result of listening to family, professional colleagues, and hundreds of students asking questions, making comments, and completing similar Activities. These class periods on special days will become learning days and looked forward to by you and your students.

May the joy of teaching be yours.

Jim Elander, retired, but still recalls the joyful teaching days.

Advisor: Olaf Elander

Three other CDs by the author:

EVERYDAY DECISION MAKING
via
GEOMETRY ESSENTIALS

A CD incorporating decision making skills via teaching the essentials of 2-D and 3-D Geometry as a logical system with application from the real world,

EVERYDAY DECISION MAKING
via
MATHEMATICAL BRIDGES FOR YOUR FUTURE

A CD course designed for the Liberal Arts or non-science majors who have completed Algebra 1 and Geometry and plan to continue their formal education, plus improve their Decision Making skills.
(This course is more thorough than the Career course.)

EVERYDAY DECISION MAKING
via
MATHEMATICS FOR A BETTER CAREER

A CD for students who have completed Algebra 1 and Geometry and need another year of math to improve their basic understandings and decision making skills. (This course is not as thorough as the BRIDGES FOR YOUR FUTURE course.)

For more information on any of the above:
Website:http://sites.google.com/site/mathfordecisionmaking/

PROLOGUE

A. Rodin's, *The Thinker*, Golden Gate State Park, San Francisco. CA.

Objective: To help students be better prepared to pass the mathematics questions on college entrance exams and other types of thinking tests by improving your Decision-Making Skills, also called Critical Thinking Skills.

THINKERS are made, not born, and scores on entrance exams are improved by reviewing the information you may have forgotten! There is an old adage that students need to understand: ***Mathematics is not a spectator sport!*** What does that mean?

An important part of becoming a CRITICAL THINKER, or a valid DECISION MAKER, besides asking questions, is to recognize words that need defining, and that all decisions are based on definitions, assumptions, and previous accepted rules or laws.

All people are eager to be THINKERS, but that ability is not a gift, it is learned, practiced and even forgotten. These activities will review some of the types and methods first encountered in geometry and some problems that are similar to the ones on the SAT, ACT and other entrance exams.

In the March 2010 issue of School Science and Mathematics, it was pointed out the two major weaknesses in college freshmen are in geometry and Critical Thinking, hence these 100 activities are involving direct and indirect thinking skills, forms of implications and other types all designed to convert those difficult special days into unique learning fun days.

These activities are designed for students (working in small groups) to create the learning situations by doing them with the teacher guidance assisting them to the conclusions.

Many of my former students commented that what they enjoyed and learned from the classes, more than anything else, was activities like these, hence, my reason to share some of these with other teachers.

Jim Elander,

(Former high school and college teacher, plus being an author, now retired in beautiful Missoula, Montana)

Email: elanderje@gmail.com

(A bit older looking now and a better golfer)

CHAPTER 1

A. Rodin's, *The Thinker*, Golden Gate State Park, San Francisco. CA.

Statements are true or false, Conclusions are Valid or Invalid?

A set of activities to help make the days like the ones before vacations or school events, not only fun days, but also learning situations. How is this accomplished? The answer is to use problems that involve all students in activities that unknowingly expose the students to a learning situation.

These activities were acquired by reading interesting books (see bibliography), attending professional meetings and testing these types of interesting problems in the classroom. The results impressed the author and he recognized their potential value, hence the desire to share them with others. Many former students indicated this type of activity was what they really remembered, enjoyed and learned from the class.

Why the following activities? Many time people make conclusions based on what they see. It is important the students realize that people see things differently and consequently the conclusions are different.

They should also be aware of the weakness in arriving at a conclusion from Inductive Reasoning. It is important that students understand that all conclusions are based on the interpretations of un-defined terms, defined terms, assumptions and previous conclusions, often called laws (in the Math World they are called Theorems).

Activity 1. Predictions may be Wrong!
            Inductive Reasoning is used, it is safe to say, by all people. The following case will illustrate the weakness in this type of reasoning.

      a. Draw a circle. (Use a compass or trace a round container, the larger the better.)

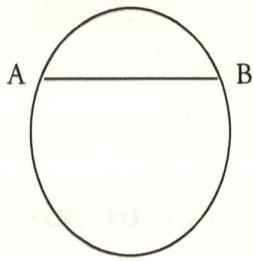

      b. Complete the following table.

Points	Chords	Regions	
2	1	2	Shown above
3	3	4	
4	?	?	
5	?	?	
6	?	?	Record the number you think is the number of regions and then actually count the number.

Answer: 31 regions instead of the 32 the students predicted.

Activity 2. One Sided Paper or the Mobius Belt

    a. Take an 8 by 11 sheet of paper and cut a 1 by 11 sheet and label is as shown below.

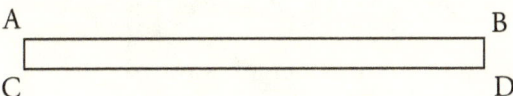

    b. Cut 3 similar strips to the one above and label the same way.
    c. With one of the strips fold it to form a belt-like strip so that B is on A and D is on C. Ask the class what they think will result if you cut the belt-like strip in half Lengthwise? Then actually cut it. What do you have?
    d. Take the other two strips and fasten the belt-like strip with tape so that B is on C and D is on A. Then cut the belt-like strip in half lengthwise. What do you have?
    e. On the third or last strip, fasten it the same way as in "d" but instead of cutting it color one side red and the other side black. What is your conclusion?

Comment: Ask the girls to bring scissors to class for the problem and make this a class project with students woking in boys-girl pairs.

The Mobius Strip was invented in 1800's by A. Moebius (1790-1868) and the concept was used in the ribbons for computer printers in the 1980's. Why? Both "sides"(if there are two sides!) were used instead of just one side on the ordinary belt ribbon. Ribbons arc not used anymore in printers since most people want color.

Report: A.F. Mobius and his contribution to the world of Mathematics. The name is spelled several ways.

Activity 3. Taxicab Geometry
To the taxicab driver the 2 points-one line concept is not always valid. The illustration below is a simple example. To go from A to B there are 2 possible routes. The cab is **always** moving in the direction of B and remains on the streets.

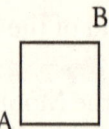

Conclusion: In a 1 by 1 block situation there are 2 ways for the cab
               driver to go from A to B.

               Determine the number of ways to go from A to B in the
               following cases by counting.

               Case 1.   1 x 2

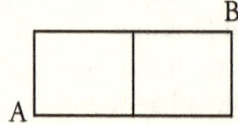

               Case 2.   1x3

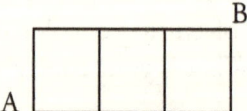

Predict the number of ways in a 2x3 case and then count the ways.

               Case 3.   2 x 3

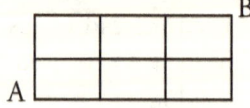

               Answers.   Case 1. 3   Case 2. 4   Case 3. 10

There are formulas for calculating the number, but they are quite difficult.

Activity 4. Polygon Angle Sum

If a triangle contains 180 degrees as you proved in geometry then how many degrees are in the sum of the angles in a N-gon? Complete the following table until you see the relationship between the number of sides and the sum of the angles and then draw a conclusion for the sum the angles in an N-gon.

The N-gons   Sum of angles in degrees

Sides	Name		
3	Triangle	180	
4	Quadrilateral	360	
5	?	?	
6	?	?	
7	?	?	
8	?	?	
9	?	?	
10	?	?	
n	?	?	

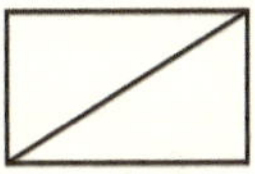

Answer: $D = (n-2)\, 180$

Activity 5. Curves from Straight Lines (Teacher should do this Activity prior to the class in order to predict the problems and questions the students may have.)

a. On a sheet of 8.5 x 11 of paper draw the following figures. (Fold it in half length wise and use both sides) Use a ruler to draw the figures and make them as large as possible. Number the length of the segments using centimeters as shown in figure 1.

b. Draw each ray or segment 12 centimeters long on a half
sheet of paper for figures 1, 2, and 3. Then number from
the each vertex (vertices) the 12cms on each ray as in figure
1 and connect the number 1-12, 2-11, etc and observe the
curves formed by lines. The numbering is more complex
for figures 2 and 3. (Practice at home)

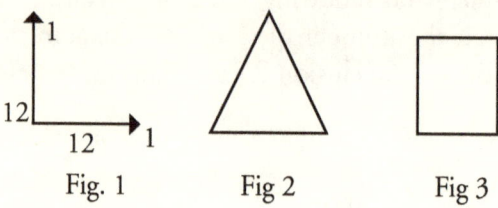

Fig. 1   Fig 2   Fig 3

c. Project: Create your own design.

Suggestion: If a few students seem to really enjoy this form of art, suggest
they create their own unique art using colored thread and
cardboard or wood.

The results for figures 2 and 3 are shown below.

Figure 3	Figure 2

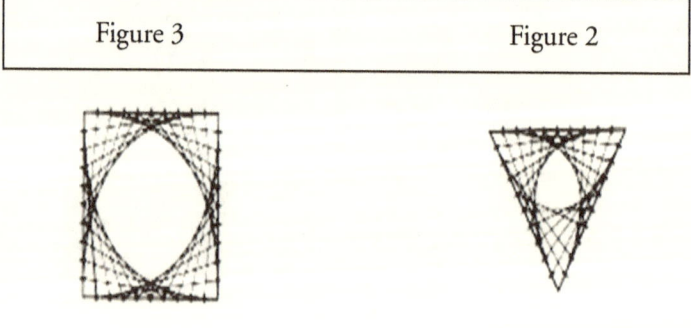

Activity 6. What do you see? Illusions

Many decisions are base on optical illusions. Example: Witnesses can testify as to what they saw at the scene of an accident. What do you see in the following?

Case 1.

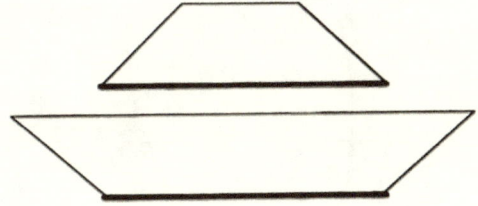

Which of the dark line segments appear longer or are they the same measure?

Case 2.

In the figures below you may see different situations as you look at the figure and blink you may see a box inside or outside a larger box.

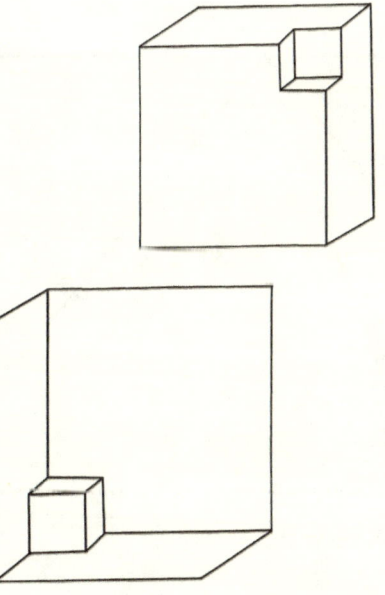

Case 3.

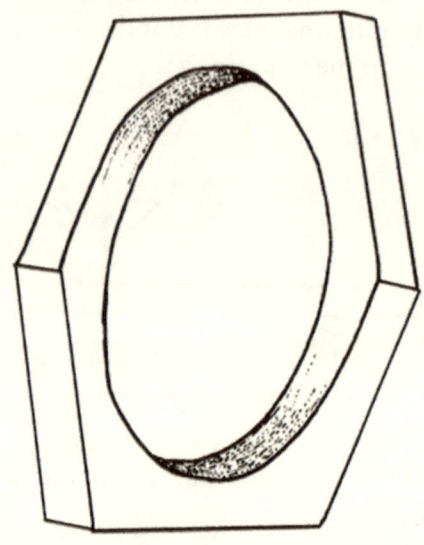

Activity 7. More Illusions

Case 1. Which of the two segments appears longer in the following figures, A or B?

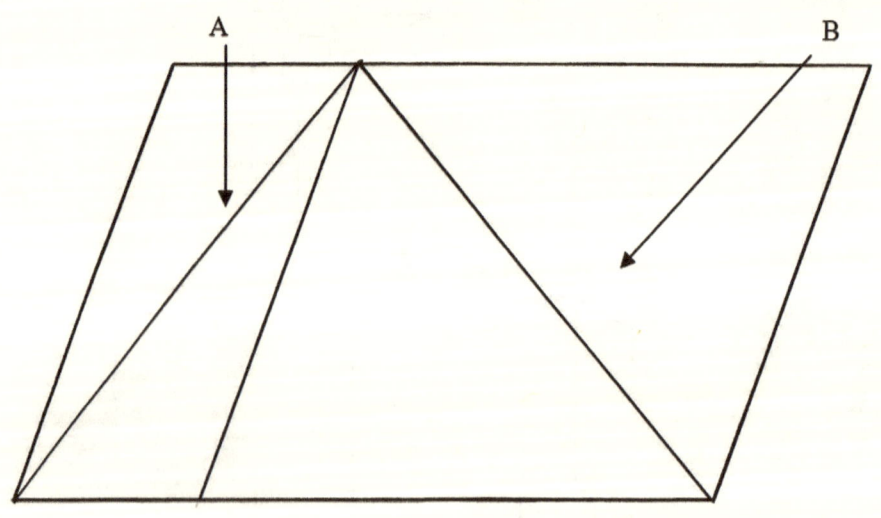

Measure them. Were you correct?

Case 2. Figure 2 was assembled from figure 1. Calculate the area of each figure and compare the answers. Surprised?

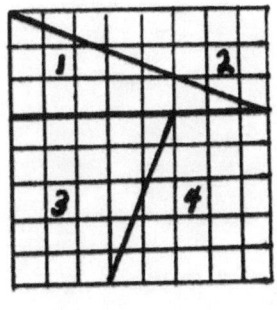

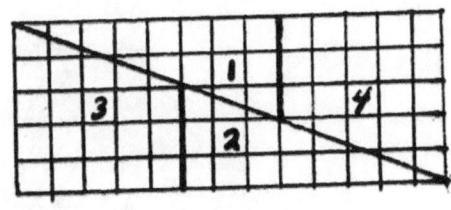

Figure I

Figure II

Case 3. Could you build this box? ?

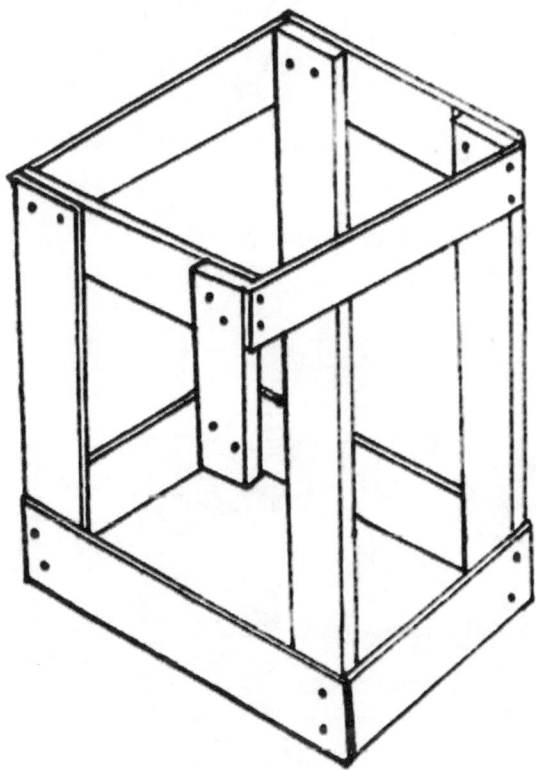

Case 4. This illusion has been used in many different ways.

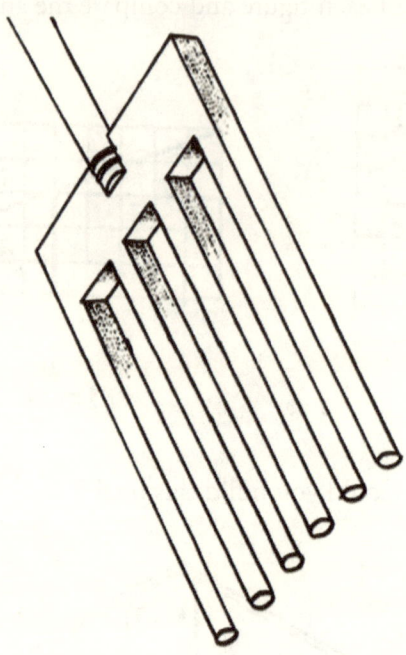

Activity 8. Teen or Sr. Citizen

  Look at the picture below, first from the left side and then from the right side. What do you see?

Activity 9. What do you see in the following pictures?

A.

B.

Suggestion: Poll the class as to what they see and draw a conclusion.
(Frog and/or Horse)

Activity 10. How many times can a piece of paper be folded in half?
Ask for predictions.
User different size sheets and different types of paper and let
the students make conclusions from their observations and
results.

Activity 11. Looking for the easy way

$5 \times 5 = 25$
$25 \times 25 = 625$
$35 \times 35 = ?$
$45 \times 45 = ?$
$55 \times 55 = ?$
$65 \times 65 = ?$

Continue the table each time predicting the answer, checking your answer with your calculator until you see the easy way to get the answer. Then write the answer for the next problem.

$(N5)^2 = ?$

Answer: N as digit 1, the N +1 as digit 2, then add on the number 25.

Activity 12. Why cars and boats tip over!

Case 1.
   Step 1. Draw an equilateral triangle on cardboard about 8 inches per side. Cut out the triangle.
   Step 2. On a sheet, say 12 by 12 inches. locate the approximate center and erect a perpendicular.
   Step 3. Label the triangle ABC from step 1, clockwise, with A at the top vertex and place the triangle so that B is at the foot of the perpendicular. See the diagram below.

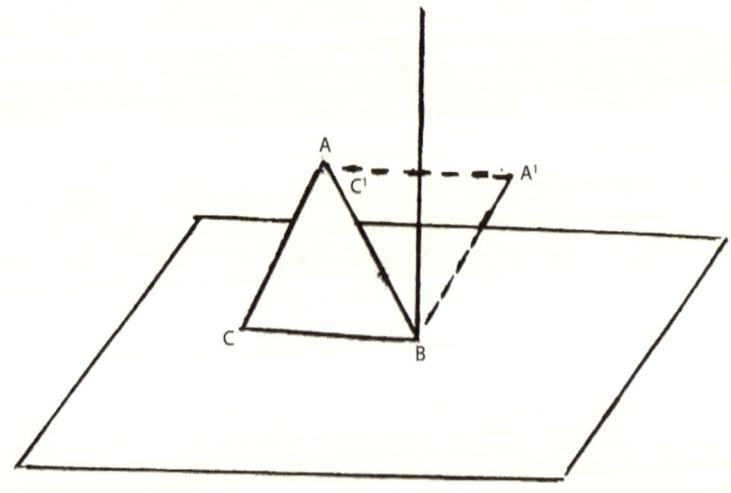

   Step 3. With B the rotation point, rotate the triangle (which is perpendicular to the 12 by 12 Sheet) clockwise to the right over point B slowly and when it begins to fall by it self, mark a point on the

triangle (dotted triangle)where the perpendicular intersects the side CA.

Step 4. Draw the line segment on the triangle determined by B and the point determined in step 3.

Step 5. Repeat steps 2 and 3 with A as the rotation point.

Step 6. Where the two line segments intersect is the center of gravity and in this case the tipping point. (For a triangle, it is the intersection of the medians.)

Case 2. Repeat the steps in case 1 using a quadrilateral, which is similar to a model car case. The concept is the same. This will locate the center of gravity or the tipping point for the quadrilateral or car.

Comment: The above cases will be remembered and talked about by the students and help their understanding as to when a car or boat tips. Perhaps a student would like to determine the tipping point of a model car.

Activity 13. All circles have the same measure for the circumference. Your kidding!

The is an interesting locus problem. Ask for their predictions before showing the answer. A smaller wheel is bolted to the larger so that as the larger wheel makes one complete revolution, then the smaller wheel does also (See the figure below). Since the distance A to A and B to B are the same then the two circumferences are the same. Hence all circles have the same circumference! Think about that! Do you understand the figure and the reasoning?

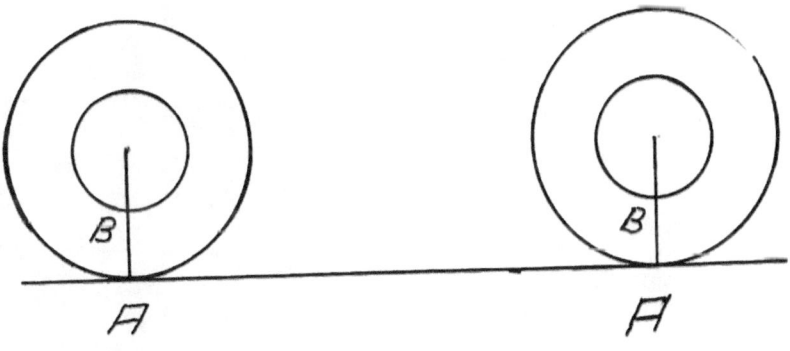

Now you know something is wrong with that conclusion, but what is it??

This can be done with the two wheels on the chalk board (white or black) or pass-out cardboard wheels and let each group perform the following. With the two concentric wheels bolted together put holes at A and B. Insert a pencil in each and carefully roll the large wheel making the path of A in one revolution and the pencil in B will trace out the path of B. The class will be surprised at their discovery. (The curve the point B makes is the cycloid. Opportunity for an interesting report by a group of students.)

The teacher should work this in advance of the class in order to answer the student's questions.

Activity 14.　FedEx Logo
　　　　　　　Do you see the arrow on every FedEx truck? You will now!

FedEx

Hint:　Focus your eyes on the blank space between the lower half of the large E and the small X. Once you see it you will never forget it and will see it every time you see a FedEx truck!

**Mathematics uncovers the "mystery"
behind a problem.**

CHAPTER 2

A A. RODIN'S, THE THINKER, Golden State Park, San Francisco, CA

Number Theory—Algebra—Reasoning
Activities Potpourri

A program to make the days like the ones before vacations or a eventful weekends, not only a fun day, but also a learning situation. How is this accomplished? The answer is to involve all students in activities that unknowingly expose the students to a learning situation. Let the students explain their solutions and answers.

These activities were acquired by reading interesting books (see bibliography), attending professional meetings and trying these types of interesting activities in the classroom. The results impressed the author and he recognized their potential value, hence the desire to share them with others. Many former students responded positively as to the recollection of these.

Activity 1:   Which areas of study are important for the future?

You know that students always asking why "They need to know a subject or what will the benefit be?"

This Activity will help a student determine which subject will help the most in preparation for college or vocational school. (If it were only this easy.)

Suggest each student select the number corresponding to their guess, related to the most beneficial subject.

1. English
2. World History
3. Biology
4. Physics
5. U.S. History
6. Chemistry
7. Athletics
8. Activities
9. Mathematics

After their selection as to their guess, then inform them to perform the following steps that will reveal the area that provides the best preparation.

1. Select your guess as to the subject area and multiply the corresponding counting number by 3.
2. Add 3 to your answer in step 1.
3. Multiply the number from step 2 by 9.
4. Add the digits in the answer to step 3 until you have a one-digit number.
5. Match your answer to step 4 with the corresponding subject or program. This is the subject area most valuable for you.

Comment: The answer will always come out 9, indicating Math is the most helpful! This answer is prearranged based on what is called Casting out Nines. Which was taught at one time in the schools before calculators and should now be used in checking solutions to addition, subtraction, multiplication, or division problems in the grade schools.

Example: 7539 + 8642 = 16181 Now add the digits in the three numbers resulting in 24, 20, and 17. Now add the 20 and 24, which is 44. Now add the two digits of 44 to give the single digit 8. Do the same with 17. (The sum of the digits in the answer.) and the sum is 8. Since the two numbers (7539 and 8642) which were added give 8 and the original answer (16181) also gives 8 you know the answer 16181 is correct. (This method of checking works for the four operations.)

Activity 2:   This problem will tell the student if their answer is correct. The way the teacher introduces a problem to the students creates an intriguing problem and generates the mystery as to how a problem can tell the right or wrong solution.

1. Instruct the students to write the first 9 counting numbers and circle the one they write the poorest.
2. Multiply the circled digit by 9.
3. Multiply the answer in step 3 by the following, 12345679. (no calculator)
4. The correct answer will contain only the circled digit. If the student doesn't know their answer is correct, then suggest they do the multiplication again.

Activity 3:   How to become a millionaire in 30 days. Again the way the teacher introduces the activity creates the Interest. This problem can create the curiosity for the class simply by the title.

Method: Give each student a 30 day month calendar and have the students mark in the first day square to indicate one penny, Also in square 1 put a 1 to indicate the sum amount. The square for day 2 should contain a 2 (for 2 cents) and a 3 for the sum. Let the students complete the calendar using a similar method for the 30 days. (It is important that they learn to create "tables to organize their data" and draw conclusions from the data.) Some students will write the amounts as powers of 2, so on the 5th day square they deposited 2^4 pennies and the sum is 2^5-1.

Comment: Each deposit is a power of 2
Each sum is a power of 2^n - 1
Then arrive at the total amount for the 30days, hence a millionaire.
(The 30 day sum is $10,737,418.23.)

Activity 4.   How large is the college class?
Many times in college the lecture class size is larger than in high school.
Professor IQ reported 1/5 of his students earned A's, ¼ earned B's, ½ earned C's, and 3 students failed. How large was the lecture class?

Answer: 60

Activity 5.   How many of each were purchased?
A teacher purchased the following for use in her classes. Colored magic markers at $5 each, pens at $1 each, and pencils at 10 for a dollar. The total number of items was 100 and oddly the total cost was $100. How many of each were purchased?

Answer: 9 markers, 51 pens, and 40 pencils

Comment: If you try to solve this using algebra, it turns out to be a Diaphontine Equation. These equations have more variables than equations, but the other condition or restriction is the requirement of counting numbers for answers.
Ask the class for some conditions, like the number of markers is less than what number? This will narrow down the possibilities.
These conditions are: (Let the students list additional ones.)
         Counting numbers only
         There are less than 100 of each item.

Activity 6. What number?

Henry challenged the class with the following problem. "I am thinking of a counting number less than 100 and if you divide by 3, or 4, or 2 the remainder is 1, but if you divide by 5 the remainder is 0. What are the numbers?" One student said the number ends in a ? or ? What did the student know and understand?

Answers: 25 or 85

Activity 7. Dad, please send money!

A college student, as you would expect, often sends a note home asking for more money. This student knew his parents enjoyed little tricks or puzzles, so he emailed the following puzzle request. Using only the ten digits representing letters he sent the following message. (Only one digit for a letter, example if *E* is 5 then no other letter can be 5, but if there are two *E*'s then both would be 5.) The message was:

$$\begin{array}{r} \text{SEND} \\ + \underline{\text{MORE}} \\ \text{MONEY} \end{array}$$

Comment: Ask these questions for explanations

Why does M have to be 1?

Why does S have to be an 8 or 9?

The father placed the decimal point in the answer between the N and E. Why?

How much money will the father send?

Give the groups time to solve the problem and then explain their solution.

Answer: 95.67   Students may have a different answer.
         + 10.85
         $106.52

Activity 8. Clock Arithmetic

There are to cases to be considered, the 12-hour clock or the 24 hour clock.

Case 1. The 12-hour clocks which requires a.m. or p.m. to tell
the time of day.

Case 1. 1. 2 + 4 =?   2. 3+ 6 =?   3. 4+ 6 = ?   4. 6+ 8 = ?
Answers: 6   9   10   2

                5. 8 - 9 = ?   6. 10 + 11 = ?   7. 3 - 8 = ?   8. 3 x 8 = ?
Answers: 11   9   5   ?

Case 2. The 24-hour clock (Military time)

Case 1. 1. 2 + 4 =? 2. 3+16 =? 3. 14+16 = ? 4. 16+ 23 = ?
Answers: 6   19   6   39 = 15

                5. 8 - 9 = ? 6. 10 + 11 = ? 7. 3 - 8 = ? 8. 3 x 9 = ?
Answers: 23   21   19   ?

Suggestion: It helps to have a large clock (can be made out of cardboard)
of each type to demonstrate the answers. Make up some more
problems for the groups to solve.

Activity 9. 1=0?? How can this be?
The symbol (∞) in the following represents a very large number.

Case 1. 1/∞ = 0 or 1 = 0 (∞) or 0, which now reads 1 = 0.

Case 2. ∞/1 = 0 or ∞ = 1(0) or 0, which now reads a very large
number equals 0 How can this be?

Activity 10. Average speed (SAT or ACT question)
A person average 20 mph driving to work in the rush hour.
Leaves early and averages 30 mph diving home. What is the
average MPH for the round trip?

Answer: a.   25 mph, the average of 20 and 30.
             b.   23.6 mph since it took longer at 20 mph
             c.   24.6 mph since it took less than 25.
             d.   26.4 mph since the home trip is faster.
             e.   24mph

Comment: Poll the class for their answers and then explain that the correct answer is 24 mph.

Many will select the wrong answer and their curiosity as to why will be amazing.

Solution: Average speed is D/T or and the distance is D (one way), then the average is Total distance divided by total time or D/20 +D/30.

$$\text{Ave. mph} = 2D/(D/20 +D/30) \text{ or } 24\text{mph}$$

Activity 11. Summer pay

Pete and John both had summer jobs for June, July, and August that paid the same amount the first month. But at the end of June, Pete was given a 10% raise and John had to take a decrease of 10%. Ironically, on August 1st Pete was given a 10% cut and John was given a 10% increase.

Questions
  1. What did each student make during the summer, if the June pay was $100??
  2. What did each student make in the month of August?
  3. What did each student make during the summer, if the pay was $P?

Answers:
  1. Pete: 100 + 110 + 99 = $309 John: 100 + 90 + 99 = $289
  2. Pete made $99 and John made $99.
  3. Pete made $3.09P and John made $2.89P.

Activity 12. Diagonal of a 4th dimensional cube from a mathematical viewpoint.

  a. Draw a square of side s and calculate the length of the diagonal.
     Answer: D of a square is $D = s\sqrt{2}$.
  b. Draw a cube of side of s and calculate the length of the cube's diagonal.
     Answer: $s\sqrt{3}$

   c. By inductive reasoning the diagonal of a 4th dimensional cube is s√4 or 2s.
   d. Predict the length of the diagonal of a 5th dimensional cube.

Comment: Ask a student to look up the word Tesseract.

Activity 13. How many lanes is the coach planning for?

The college track coach suggested to the budget committee that the college build a mile circular track and add 200feet to the circumference for the width of the track. The committee doubted the 200feet would be enough.

What do you think? In the drawing below the track is the gray area.

If a lane is about 3 feet wide, then how many lanes is the coach planning for?

The answer is one of the following, 2, 4, 6, 8, 10, 12. (Justify your answer.)

Justification: $2\pi RL - 2\pi RT = 200$

$$RL - RT = \frac{200}{2\pi}$$

Track width is 31.8 feet or 10 full lanes.

Activity 14. Saturday's pay

Three high school freshmen sold ice cream bars at Saturdays Game and decided to share the tips based on how many bars each sold. The pay per day was $10 plus tips. The total tips came to $14.95. If Joe sold 2/3 as many as Art and Fred sold ¾ as many as Art, then how much did each earn in tips?

Answers. Art $6.19, Joe $4.13, Fred $4.65

**People understand the inductive fallacy, but still "jump"
to the questionable conclusion.**

CHAPTER 3

Rodin's THE THINKER, Golden State Park, San Francisco, CA

Inductive Reasoning conclusions
and related Problems

A program to make the days like the ones before Homecoming, not only
fun days, but also learning days, instead of wasted learning days. How is
this accomplished? The answer is to use problems that involve all students
in activities that unknowingly expose the students to a learning situation.
Conclusions reached by inductions means the conclusion is arrived at by
observing a series of events and then assuming the conclusion will always
occur. Many old sayings were based on inductive reasoning such as *Red skies
at night is sailor's delight.* The weakness is that the conclusion predicting
future events is based on passed events.

Definitions:

Inductive Reasoning is arriving at a general conclusion from a few cases.

If you are in circle A, then you are in B. (Not necessarily valid)

Deductive Reasoning is arriving at a conclusion from a set of definitions, assumptions and possibly previous valid conclusions.

If you are in circle B, then you are in circle A (valid)

You will probably have to think about these definitions a while to really understand each one and be able to apply them. The following activities will help.(you may wish to use the circle diagram to explain the relationships of the converse, inverse and the contrapositive to the original statement

The ultimate objective in mathematics is to improve ones decision-making skills. All decisions are based on **undefined terms**, **defined terms**, **assumptions**, and previous accepted decisions (laws or theorems in mathematics). Plato recognized this and it was that reason the following was posted at the entrance to his academy. *Let no man ignorant of Geometry enter here.* One of the greatest examples of this is Jefferson's Declaration of Independence. It was B. Peirce, a mathematician in the 1800s, who defined mathematics as the "science of drawing necessary conclusions." One other point for you to think about is: Statements are true or false, conclusions are valid or invalid.

This program can make the days like the ones before Homecoming, not only fun days, but also learning situations, instead of wasted learning days. How is this accomplished? The answer is to use problems that involve all students in activities that unknowingly expose the students to a learning situation. Conclusions reached by induction means the conclusion is arrived at by observing a series of events and then assuming the conclusion will always occur in the same way in the future. Another old saying based on inductive reasoning is *Early to rise makes one healthy, wealthy and wise.* The weakness in this type of reasoning is that the conclusion predicting future events is based on passed events no matter how few.

The unique way the problems are introduced creates the interest and decreases or eliminates problem days before vacations, holidays or special events. Eventually the students will look forward to the next activity day.

Activity 1. In the above paragraphs, the statement that all decisions are based on undefined terms, defined terms, assumptions, and previous accepted decisions, probably created questions. A wonderful example of this is Jefferson's
Declaration of Independence.
The following will illustrate the use of undefined and defined terms.
Classify each word in the **Preamble** to the **U. S. Constitution** (below) as un-defined or defined. This will illustrate the point.

> *We the people of the United States, in order to form a more perfect Union, establish Justice, insure domestic Tranquillity, provide for the common defense, promote the general Welfare, and secure the Blessings of Liberty to ourselves and our posterity, do ordain and establish this Constitution for the United States of America.*

Undefined _____
Defined   _____
Total   _____

What percentage of the total number of words are un-definable? (About 38% but answers will vary?)

Activity 2. Inductive reasoning is used by all people! This Activity will illustrate the weakness in this type of reasoning.

a. Draw a circle. (Use a compass to draw a circle or trace a round container, the larger the better.)

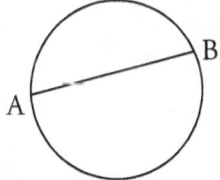

b. Complete:

Points	Regions	
2	2	The circle is divided into two sections.
3	?	How many sections?
4	?	Your prediction? Then count the sections.
5	?	Record your prediction. Then count.
6	?	Record your prediction for the number of sections and then actually count the number.

Answer: 31 regions instead of the 32 that students will predict.

Activity 3. Points and Lines

In geometry and in everyday occurrences people know that 2 points determine one line. How many line segments are determined by n non-collinear points? Start with 2 points, then 3 points and create a table matching points and segments. Look for a relationship between points and the number of lines. If there is one, then write conclusion in if-then form. What does "non-collinear" mean?

Points (n)	Lines
2	1
3	?
4	?
5	?
6	?

Continue the table and write the inductive conclusion for an easy way to predict the number of lines. (Why is this an inductive conclusion.)

Answer: Table:

Points	Lines
2	1
3	3
4	6
5	10
n	$n(n-1)/2$

Comment: This result is by inductive reasoning, but the general conclusion can be proven by Math Induction. If the class is Advanced Algebra, you may want to play professor and present the proof the first day back.

Activity 4. In taxicab geometry the above conclusion is not valid. The assumption that two points determine one line is not always true. The diagram below illustrates that fact and shows there are two routes from A to B for the cab. The rule is that the cab must stay on the segments (streets) and always be moving in the general direction of B.

Case 1.   1 x 1

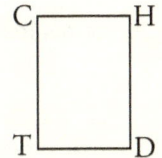

Conclusion: In a 1 by 1 block situation there are 2 ways for the cab driver to go from T to H.(TCH or TDH)
How many routes in each of the following cases?

Case 2.   2 x 1

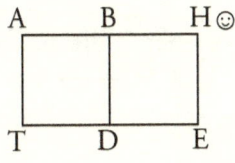

How many routes are there for the Taxi (T) to take to get to H?

Case 3.   2 x 2

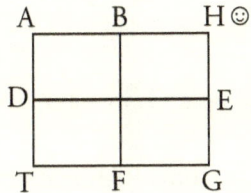

How many routes are there for the Taxi (T) to take to get to get to H?

Case 4.　3 x 3

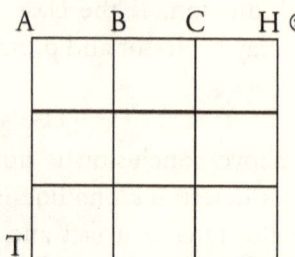

A　　B　　C　　H ☺

How many routes from T to H for the taxi? You may wish to label the un-marked vertices.

T

Answers.　Case 1. 2　Case 2. 3　Case 3. 6　Case 4. 21

Comment: Add additional R x H routes for the groups to travel. Is their a pattern?

Activity 5.　N-gon and diagonals

In Activity 3 above you arrived at the inductive conclusion that the number of lines determined by n non-collinear points is n(n-2)/2.

Complete the following table for the number of diagonals in a concave polygon for the number of sides indicated. Hint: You may need to draw each polygon and count the diagonals.

Sides(n)	Number of diagonals
4	2　⟶　⧖
5	?
6	?
7	?
.	.
.	.
n-gon	?

Answer:　Number of Diagonals in a(Pn) = n(n-3)/2

Comment: This is inductive reasoning, but the formula can be proven by math Induction. Play professor in the next class and prove the formula.

Activity 6. The Tower of Hanior Problem. This is a good problem to have the students work in groups of 2 or 3 and record their results. The original problem consisted of the following board and a set of rings of different sizes. Put a large ring on post #1 and a smaller ring on top of the large ring. The objective is to put the rings on one of the other posts in the same order by moving only one ring at a time and never having a large ring on top of a smaller ring. Keep track of the number of moves.

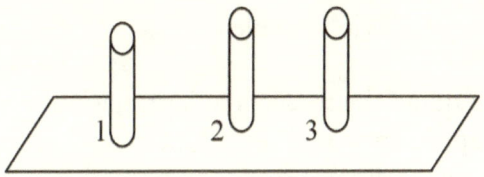

Complete the following table.

Rings	Moves	
2	3	
3	?	
4	?	
5	?	
.		
.		
.		
N	?	They may observe a pattern to predict the number of moves for N coins

Note.   Use coins instead of rings (Half dollars, quarters, nickels, dimes, and pennies.) and just mark X's on the paper for the posts.

Activity 7. 10 Coin Problem

Suggestion: Have the students work groups of 2 or 3 to solve this problem.
Arrange 10 coins as indicated below.

☺ ☺ ☺ ☺ ☺ ☺ ☺ ☺ ☺ ☺
1 2 3 4 5 6 7 8 9 10

Objective is to end up with 5 piles of 2 coins each by the following rules.
1. A move consists of jumping 2 coins and landing on the third.
   Example: Coin 1 could jump coins 2 and 3 and land on coin 4.
2. Once a pile consists of 2 coins it is "dead" or can't be moved, but can be jumped.
3. A coin can jump a pile of 2 coins and land on a single coin.
   Example: In the case 1 above coin 3 could jump the 2 coins (1-4) and land on coin 5. This results in coin 2 being impossible to move since it would have to jump 4 coins which violates rule # 1.

Note: This requires the students to keep track of their moves so they can explain their solution to the class. This is a good method and habit to develop for future research.

Answer: Coins numbers and moves
6-9, 4-1, 8-3,2-5, 7-10

Activity 8. Cook's problem:
The camp cook needed 4quarts of water to make orange juice for the campers, but he only had 3, 5 and 8 quart containers. He asked a geometry student for help. The student experimented a while and then solve the problem using only the three containers and the 8 qt container full of water.

Moves	8	5	3
1	8	0	0
2			
3			
?			
?			

Record your moves and explain your solution to the class.

Answer.

Moves	8	5	3	Containers
1	8	0	0	
2	3	5	0	
3	3	2	3	
4	6	2	0	
5	6	0	2	

6	1	5	2
7	1	4	3
8	4	4	0

Note: Students may have a different order, so let them explain their solution to the class. Praise them for their thinking!

Activity 9. Given 4 fours (4,4,4,4) try to write the first 20 counting numbers using any operations or combination of symbols.

Examples: 1. $(4+4)/(4+4)$ is 1
2. $4(\sqrt4)/[(\sqrt4)(\sqrt4)]$ is 2

Note: Keep track of their solutions.

Some answers are:

3 is $(4 + 4+4)/3$	4 is $4\sqrt4 - \sqrt4 - \sqrt4$
5 is $\sqrt4 + \sqrt4 +4/4$	6 is $4 + 4 - 4/ \sqrt4$
7 is $4 + 4 - 4/4$	8 is $4 +4 / (4/4)$
9 is $4 + 4 + 4/4$	10 is $4 +4 +4 - \sqrt4$
11 is $44/(\sqrt4 + \sqrt4)$	12 is $4 +4 + \sqrt4 + \sqrt4$
13 is $(44/4) + \sqrt4$	14 is $4 +4 + 4+\sqrt4$
15 is $4(4) - 4/4$	16 is $4 +4 + 4 +4$
17 is $44 +4/4$	18 is $4(4) + 4 - \sqrt4$
19 is $4! - 4 - 4/4$	20 is $4(4) + (\sqrt4 + \sqrt4)$

Activity 10. All of your students have played checkers (an assumption) so here is the problem.

How many squares are there on a checker Board?

Method: 1. Assume the size of a board is 8 by 8?
2. Create a table starting with a 1 by 1 board.
3. Look for a relationship between the size and the number of squares
4. Answer the question

Complete this table

Size of board	Number of new squares	Total number
1 by 1	1	1
2 by 2	4	5
3 by 3	—	—

Answer.

Size of board	Number of new squares	Total number (for an 8 by 8)
1 by 1	1	1
2 by 2	4	5
3 by 3	9	14
4 by 4	16	30
5 by 5	25	55
6 by 6	36	91
7 by 7	49	140
8 by 8	64	204

Formula: $n(n+1)(2n+1)/6$

Comment: Don't expect the students to come up with the formula, (It was derived by the method of Finite Differences.), but they should be able to complete the above table.

Activity 11. The game of Sprouts (This a variation of the game.)

(I first encountered this in a book by Martin Gardner and the game here has been adjusted for Inductive Reasoning conclusions.) It is suggested to use teams of 2 or 3 and one player keep track of the moves and outcomes to help them see the general data and predict the conclusions. Comment: This is an interesting game and will take the whole period to play and determine the conclusions.

Rules:
1. A sprout is a segment joining two seeds.
2. A seed is designated by a small circle.
Example:

Seed ⟶ ☺   ☺ ⟵ sprout joining 2 seeds

3. Only 3 sprouts can come from a seed.
4 The player to make the last move is the winner.

Sample game starting with 2 seeds.

☺   ☺

Move 1 by first player 1

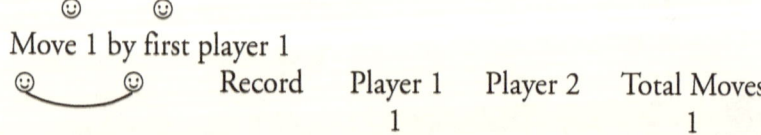

	Record	Player 1	Player 2	Total Moves
		1		1

Move 2 by player 2

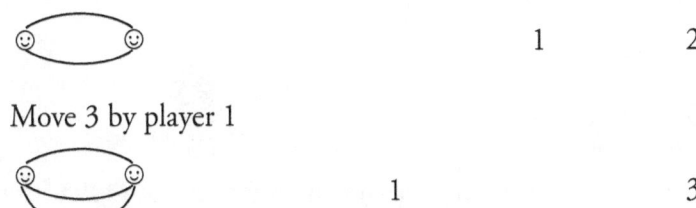

                                             1   2

Move 3 by player 1

                                       1   3

At this point the game is over and player 1 is the winner since no more move are possible due to rule number 3.

The players should record results and now play the game starting with 3 seeds.

By completing a table like the following one, players may be able to recognize some conclusions.

    Given the number of seeds, then:

        Who will win?

        The number of possible moves?

The following table is complete with results for the 2 sprouts game above.

Number of seeds to begin with S = ?		Moves (Player 1)	(Player 2)	Winner 1 or 2	Moves Total
2		2	1	1	3
3	Predict first	?	?	?	?
4	"	?	?	?	?
5	"	?	?	?	?
Additional games					
If S is odd		?	?	?	?
If S is even		?	?	?	?

Activity 12. What is wrong with this conclusion?

        36 in. = 1 yard Valid

        9 in. = ¼ yard Valid

        3 in. = ½ yard Invalid Why?

Activity 13. Interesting quotients

        Using your calculator record the answers to the following.

        1/9 = ?

2/9 = ?
3/9 = ?
4/9 = ?
5/9 = ?
6/9 = ?
7/9 = ?
8/9 = Predict your answer and check it with your calculator.
9/9 = Predict your answer and check with your calculator.

Activity 14. The Way Out.

Many carnivals or game parks have a house or a maze where you enter and try to find the path to the exit. The following few examples lead the student an inductive general conclusion.

Case 1. If you connect points A and B, the segment crosses one boundary segment and you can see A and B are on different sides.

Case 2. If you connect points A and B, the segment crosses boundary segment twice and you can see A and B are on the same side

Case 3. If you connect points A and B, the segment crosses the boundary segment three times and you can see A and B are on the opposites sides.

General Conclusion. If the maze is closed and separated into two regions, then the line segment joining two points will indicate the side the points are on by the number of times it crosses the dividing segments. (odd—opposite sides, even—same side)

Case 4. In the maze below:

   a.   can D visit A? B? C? D? E? or F?
   b.   can B visit C? D? E? or F?

   c. can C visit D? E? or F?
   d. can D visit E? or F?
   e. can E visit F?

Be sure the student understand the following!

1. If they are on the same side as stated in the General Conclusion, then try to find the path.
2. Looking at the figure and to find a path appears to be very difficult, but the easy way is to put your right hand (or left hand) on the "wall" and walk the path keeping hand on the wall and this will lead you to you're the exit. Remind the students that the State or County Fairs many times have a mystery event of this type.

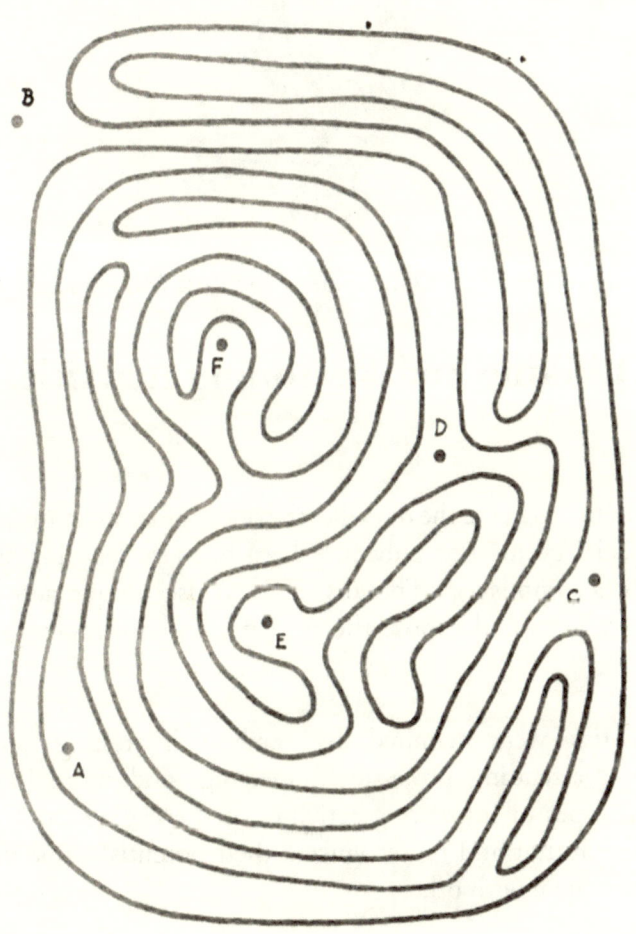

YOUNG PEOPLE WHO HAVE ACQUIRED THE ABILITY TO ANALYZE PROBLEMS, GATHER INFORMATION, PUT THE PIECES TOGETHER TO FORM TENTATIVE SOLUTIONS WILL ALWAYS BE IN DEMAND.

J. G. Maisonrouge
Board Chairman
IBM World Trade Corp.

CHAPTER 4

A. Rodin's *THE THINKER*, Golden State Park, San Francisco, CA.

Thinking and Reasoning Activities

A set of Activities to make the days like the ones before vacations, holidays, or other special events not only fun days, but also learning situations. How is this accomplished? The answer is to use unique activities that involve all students and expose the students, unknowingly, to learning situations.

These activities were acquired by reading interesting books (see bibliography), attending professional meetings and more importantly testing these types of interesting activities in the classroom. The results impressed the author and he recognized their potential value, hence the desire to share them with others.

To make the learning experience more meaningful try to adapt the problem to the community or local situations. The ultimate objective in mathematics is to improve ones decision making skills. All decisions are based on undefined terms, defined terms, assumptions, and previous accepted decisions. Plato recognized this product of geometry and it was the reason the following translation was posted at the entrance to his academy. *Let no man ignorant of Geometry enter here.* One of the greatest examples of this type of reasoning is Jefferson's Declaration of Independence. It was B. Peirce in the 1800s who defined mathematics as *"The science of drawing necessary conclusions."*

Deductive conclusions methods consist of two methods, the Direct or the Indirect. You probably use the Indirect much more than the Direct. Example of Direct. Why is $(-1)(-1) = +1$? Proof:

$-1(-1 +1)=0$ or $(-1)(-1) +(-1)1 = 0$ or $(-1)(-1) + (-1)(1) = 0$ or $(-1)(-1)+(-1) = 0$, hence $(-1)(-1)$ is $+1$ This proof uses the distributive and the identity properties.

The Indirect method consists of recognizing all the possible conclusions and showing all possibilities false but one, therefore, the last one is the valid result. You have probably seen many Sherlock Holmes mysteries on TV where he used the Indirect Method to solve the crime.

Suggestion: Work the activities in advance so you will anticipate the questions.

Activity 1. The Declaration of Independence
This is an n excellent activity for a TGIF day such as Memorial Day or on Presidents Day. Request copies of the *Declaration of Independence*, some of the local Veterans Organizations contain copies in their handouts. Then instruct the students in groups of 2 or 3 to classify contents of the first two paragraphs down to the statement (*To prove this, let Facts be submitted to a candid world.*) as to un-defined terms, defined terms, assumptions (The teacher will probably have to assist them as they get started.)

Note: If they haven't read the document suggest they finish reading it t home. You may have them count the number of undefined terms, defined terms, and assumptions. Then calculate the percentage of the total number of words are undefined. (It usually turns out to be about 40-50 %.)

Activity 2:   The President's Advisors (This Activity is good for a whole hour period. The teacher will need to do some planning in advance, like the beanies.)

The President has three advisors and needed to determine which one is the wisest.

The President could not determine who is the wisest after a discussion period on various issues with each one. So they all went a large room and the President placed each in a corner of the room. He then gave them the following instructions.

1. You will each be blindfolded and on each of you will be placed a white beanie or a black beanie.
2. When the blindfolds are removed, if you see a black beanie you are to put your hand above your head and when you know the color of your beanie you may take your hand down and declare the color of your Beanie. You will have to justify your answer, but you better be correct, or you will be dismissed.(He did this to keep them from guessing.)

The President then ordered the blindfolds removed and all the hands went up. **After a while** one advisor took his hand down and said "I have a black beanie on." How did the advisor know????

Suggestion:Let the students give their explanation, but ask for justification and eventually you will probably have to act this out, so have some black objects (hats) handy. It will take all period or more for the whole class to understand how the wise advisor knew his hat was black.

Answer: The advisor's thinking is: "My beanie is either white or black, if it were white then the other two would know immediately their beanies are black, but since they did not put their hands down mine must be black. QED. (What does QED stand for and mean?) (Quode Erat Demonstrandum)

Comment: This uses INDIRECT REASONING and it is very important the students learn this type of reasoning, since it has been stated the it is used much more then direct reasoning as taught in geometry. Some student will not understand this solution at first. To aid understanding it is suggested to actually act it out using three students.

Activity 3. The Cell Phone Problem?
Eight students whose cell phones that all looked a like but oddly one phone had a larger chip and was a bit heavier. They placed all the phones in a basket and the teacher challenge them to find the heavier one using a balance type scale only twice. How can this be done?

Answer. The 8 phones are divided into in to 3 groups (3,3,2) and weigh the two groups of 3.

Case 1. Take the 2 sets of three and place them on the scale. If one of the sets of 3 is heavier, then select 2 from the heavy set and weigh them. There are two possibilities, the 2 weighed the same (1)or they do not(2). If they weigh the same the un-weighed one is the heavy one. If they do not weigh the same, then take the one which weighs more as the heavier phone. QED (What does QED stand for?)

Case2. The two sets of three weighed the same so put the set of two on the scale, then take the one which weighs more as the heavier phone. QED.

A balance scale is like one as shown below.

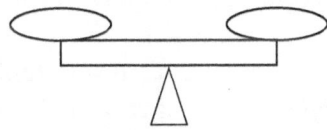

Activity 4.   Galileo and Gravity (A very good example of Indirect reasoning.)
              Galileo's (1564-1642) conclusion that objects heavier than air
              fall at the same rate. Most people believe that the conclusion was
              arrived by actually dropping objects from the Leaning Tower of
              Pisa. This would be using Inductive Reasoning and you know
              the weakness in that type of reasoning. So did Galileo! His
              deductive justification is explained below, after which he then
              went to the Tower and showed the people.

Suggestion: Ask the students as to their belief to the question: Which falls
            faster?

Explanation. His reasoning went like this. The possibilities are: (1) Heavier
             objects fall faster.
             (The popular belief of the day.) (2) Heavier objects fall slower.
             (3) They both fall at the same rate.

             Case 1:   If the heavier objects fall faster than the lighter object,
                       then tie the two objects together and the two combined
                       objects should fall faster, but if the heavier object falls
                       faster and a lighter object slower then together they
                       will fall slower than the heavier object and faster than
                       the lighter object. This is contradiction since an object
                       can't fall faster and slower at the time. (Read again and
                       think about that!)

             Case 2:   The argument is the same as in case 1, but reversed
                       and leads to a contradiction.

             Therefore, the only possibility is Case 3, they fall at the same
             rate.

Activity 5.   A new type of number
              Pythagoras (ca 550 B.C.) and his Brotherhood assumed all
              numbers could be written as fractions, such as a whole number
              over a whole number.
              A logical assumption. Can you think of any numbers that can't
              be written as a ratio of two whole numbers or integers?

Suggestion: Listen to their answers!

Then ask: Can the $\sqrt{2}$ (square root of 2) be written as a fraction? They will know the answer is between 1 and 2. (Some will give the answer from their calculator and hopefully there will be several calculators with different answers.)

Assumption 1: The $\sqrt{2}$ can be written as a fraction. The only cases are $\sqrt{2}$ = E/O or O/E. Why not E/E? (E/O means even / odd and O/E is odd / even, both reduced to simplest form) (You may have to explain simplest form.)

>   Assume (Case 1): $\sqrt{2}$ equals O/E now multiply by E and square both sides. Result is E^2 (2) = O(0). This is a contradiction or a false statement. Why? (An even number can not equal an odd number.) Therefore assumption 1 is false.

>   Assume (Case 2): $\sqrt{2}$ = E/O (E/O means even / odd) $\sqrt{2}$ equals E/O now multiply by O and square both sides 0^2 (2) = E(E) (O represents an odd number). This a contradiction.
>   Why? (0^2) = E/2(E), but an odd numbers squared results in an odd product and the E/2(E) is always an even product, therefore a contradiction and hence the Assumption 2 is false

>   Therefore the assumptions 1 and 2 are false and the square root of 2 cannot be written as a ratio of two integers and we have a new type of number, an irrational number. (Irrational meaning not a rational number.)

Activity 6.   Questionable Request and Ads
        All Conclusions are base on, un-defined terms, defined terms, assumptions and previous conclusion.

Case 1. A teacher provided this report about his club last year.

    a. At the end of the first semester we had had increased our membership 300%.

    b. During the second semester we loss only one member.

    c. We are open to all students.

       Would you approve the teacher's request to continue the club next year? The facts: There was originally, only one member and only 3 at the end of the first semester. Does this change your answer?

Case 2: It was reported that very few accidents occur at speeds over 100 mph, therefore you are safer driving at high speeds. What do you think?

Case 3. It was reported that a 100 year old woman smoked all her life, therefore smoking prolongs life. What could you conclude?

Case 4. It was reported most accidents occur within 20 miles of your home, therefore . . . (You may conclude what?)

Case 5. Pro golfers use ball O, therefore I will use ball O. What is the assumption?

Case 6. A certain cigarette ad always has a man riding a horse and smoking, what is your conclusion? (One of my students concluded: Horses like smokers.)

Activity 7. Implications and forms of Implications. (This is very important?)

Definition: An implication is a statement in the form of If-then. You encountered many of implications in Geometry and may use them in everyday situations.

Example. (Math) If a triangle is a right triangle with sides a, b and c(c is the longest side called the hypotenuse), then $a^2 + b^2 = c^2$. What is the name of this theorem?

Using Ads, like the type in Activity 6 can be an interesting activity and at times may have humorous interpretations. The forms of an implication are very valuable, will be (hopefully) remembered by the students and applied in the future for decision making.

The four forms of an implication are:

Original Implication	Symbolic form
(an if-then statement)	A → B
Converse	B → A
Inverse	~A → ~ B
Contrapositive	~B → ~ A

The important question is are these 4 related and if so how? The following diagram can be used to answer the question. This illustrates the statement: If A then B.

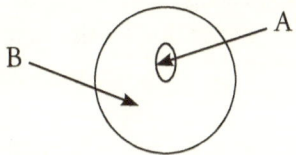

1. **Original** is A→ B then this shows if you are in A, then you are in B. True or valid
2. **Converse** is B → A, read as, if you are in B, then you are in A. This may be not true or not valid as you can see.
3. **Inverse** is ~A → ~ B which is read, if you are not in A, then you are not in B. This may be not true or not valid as you can see.
4. **Contrapositive** is ~B → ~ A which is read, if you are not in B, then you are not in A. This is true or valid.

Activity 8. Ads and implications
    Statements are True or False and conclusions are valid or invalid
    Write each of the following statements as implications and then write the 4 forms. Classify the original A→B statement as true or false and then classify the others forms.

Case 1: It was reported that very few accidents occur at speeds over 100 mph, therefore you are safer driving at high speeds high speeds.

Case 2. It was reported that a 100 year old woman smoked all her life, therefore smoking prolongs life.

Case 3. It was reported that most accidents occur within 20 miles of your home, therefore . . . (You may conclude what?)

Case 4. A pro golfers use ball O, therefore I will use ball O. What may you concluded as an assumption?

Case 5. A certain cigarette ad always has a man riding a horse and smoking, what is your conclusion?

Case 6. The following if-then statement was widely and falsely used by many during the first years of the Iraq war and many still make false conclusions.

If you wear a flag pin, then you are a patriot. This statement is probably true.

Write the converse, inverse and contrapositive of the statement.
Discuss each as to validity or invalidity. What word needs to be defined? (Validity)

Case 7. Repeat Case 6 for the following statement. ads.)

If you fly the Flag on Federal Holidays, then you appreciate the United States of America.

Case 8. Collect and bring to use in the class ads from magazines and let the class translate them for their interpretation.

Activity 9. The sum of the first N counting numbers

It has always been an interesting problem to solve for the sum of the first n counting numbers. (Even Gauss, the greatest mathematician (It is said), was given the problem for the sum of the first 100 counting numbers. It took him a very short time to solve it (He was in elementary school.). You can solve it also by Math Induction by completing the following table.

Counting Numbers	Sum
1	1
1 + 2	3
1 + 2+ 3	6
1+2+3+4	10
First 5	15
First 6	21
First 7	?

Continue the table and write your prediction, test it and then write a formula or equation to solve for the sum of the first 100 counting numbers. What was Gauss's answer for the sum?

Answer: $S = n(n+1)/2$ Gauss's answer: 5050

Suggestion: A student may elect to report how Gauss did it.

Activity 10. John's Walk

John walks the same route to and from school and passes a post office, a market, a computer store, a park and a theater, but not in that order. Going to school he passes the park before the computer store, but after the post office. He passes the theater first. Going home he passes the market second. Draw a line segment and mark H for John's and S for school at the ends. Indicate on the line segment the 5 items that John passes to and from school in their order.

Home ————————————— School

Answer: Home Theater Post Office Park Market Computer School

Activity 11. Stack Problem

You have seen displays like the following in many stores for stacking cans.

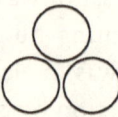

This we will call a stack with a base of 2 and it supports one can.

The problem is to determine how many cans (stack) will be supported by a base of n cans.

Complete? Base	Stack	Total cans
2	1	3
3	?	?
4	?	?
5	?	?
.		
.		
.		
n	?	?

Answer: Stack is n(n-1)/2 Total is n(n-1)/2 + N

Activity 12. One-sided paper or the Moebius Belt

a. Take an 8 by 11 sheet of paper (or larger) and cut a 1 by 11 sheet and label is as shown below.

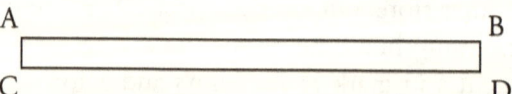

b. Cut 3 similar strips to the one above and label the same way.
c. With one of the strips fold it to form a belt-like strip so that B is on A and D is on C. Ask the class what they think will result if you cut the belt-like strip in half Lengthwise? Then actually cut it. What do they observe?

d. Take the other two strips and twist them and fasten the belt-like strip with tape so that B is on C and D is on A. Then cut the belt-like strip in half lengthwise. What do you have?

e. On the third or other strip, fasten it the same way as in "c" but instead of cutting it color one side red and the other side black. What is your conclusion?

f. As a class demonstration create another belt with a full twist (2 half twists) and cut it as in "a" after the class predicts the outcome.

g. As a class demonstration create another belt with three half twists and cut it as in "a" after the class predicts the outcome.

Comment: Ask the girls to bring scissors to class for the Moebius problem and suggest thy work in boy-girl teams.

The Mobius Strip was invented in 1800's by A. Moebius (1790-1868) and the concept was used in the ribbons for computer printers in the 1980's. Why?

Answer: Both "sides" were used instead of just one side. Ribbons are not used in most printers now. Why?

Activity 13. Who was guilty? (Indirect reasoning concept is very important.)

Hint: Use indirect reasoning. Indirect reasoning is the method that tests all the possibilities and they all lead to a contradiction but one, then that one is the correct conclusion.

A school had some windows spray painted during Halloween and the Superintendent asked the Math teacher to try to solve which of four students was guilty. The four were called in, individually, and made the following statements.

George: Chelsea did it.
Chelsea: Terri did it.
Seth: I didn't do it.
Terri: Chelsea lied when she said I did it.

If only one of these four statements is true, then who was guilty?
Answer: Seth

Activity 14. Text Message
A fellow student sent this Text message. Can you read it? Hint:
The student was an A student.

ICURICUBICUR2YS4ME

Activity 15. Magic Square
The following problem is called a Magic Square. The problem
is to place the first nine counting numbers in the small squares
so that adding up-down or diagonally the sum is always 15.

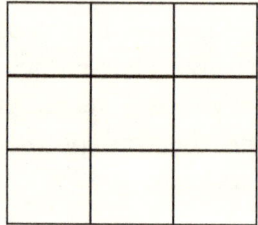

Hint: Write all the possible three number combinations that
add to 15 and see which digit should go in the center.

Answer: 816   There may be more than one answer
             357   arrangement.
             492

Activity 16. Who did it?
Another application of indirect reasoning. This is a very
important method and it has been said, that it is used the most
in everyday decisions making. Again the method is to assume
each of the possibilities and they all lead to a contradiction but
one, then that one is the solution.

Case1: A police officer questioned three students as to
the damage to a car and a pickup. Only one of the
following statements is true. Who damaged the car?
Who damaged the pickup?

Max said: "Bob did not damage the car."
Bob said: "Joe did damage the pickup."
Joe said: "Bob is lying."

Solution Method: Let's assume Bob is telling the truth, then:
Joe did damage the pickup
Joe did not damage the pickup.
The assumption led to a contradiction, therefore the assumed statement is false.

Now assume Joe is telling the truth and what is your conclusion? If Joe isn't telling the truth, then Max must be telling the truth. Then assume Max is telling the truth.

Answer: Bob damaged the car and Joe damaged the pickup.

Activity 17. Clock arithmetic.
Making Clock Arithmetic understandable and easy. There are two cases to be considered, the 12-hour clock or the 24 hour clock

Case 1. The 12 hour clocks which requires a.m. or p.m. to tell the time of day.

Problems.	1. 2 + 4 =?	2. 3+ 6 =?	3. 4+ 6 =?	4. 6+ 8 =?
Answers:	6	9	10	2

	5. 8 - 9 =?	6. 10 + 11 =	7. 3 - 8 =	8. 3 x 8 =?
Answers:	11	9	5	2

Case 2. The 24 hour clock (Military time)

Problems.	1. 2 + 4 =?	2. 3+16 =?	3. 14+16 =?	4. 16+ 23 =
Answers:	6	19	6	39 = 15

	5. 8 - 9 =?	6. 10 + 11 =?	7. 3 - 8 = ?	8. 3 x 9 =?
Answers:	23	21	19	3

Suggestion: It is suggested to have a large clock (Cardboard) of each type to illustrated or demonstrate the answers.

Activity 18. The eye deceives you.

Many decisions are base on optical illusions. Example: Witnesses testify what they saw at the scene of an accident and can contradict each other.

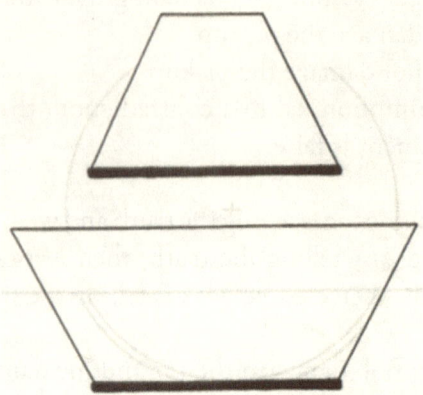

a. Which of the dark line segments appear longer or are they the same measure?

b. What do you see in the following 2 pictures? In the first one. Do you see a young woman or an old woman? (I have shown this to hundreds of students and have concluded more left handers see the young woman than right handers.) In the second one, a horse or a frog?

Do you see a horse or a frog?

c. In the figure below, do you see 3 cubes or 5? Now rotate the
    figure 180 degrees and answer the question again. Witnesses
    can report different conclusions!

Most will reply 3, but invert picture (or stand on your head) and you will see 5.

Activity 19. Proof that 2 equals 1 or does it?
    If $x = a$, then multiplying by x gives $x^2 - xa$ and subtracting
    a^2 gives $x^2 - a^2 = xa - a^2$, therefore $(x - a)(x + a) = a(x - a)$ and
    dividing by $(x - a)$ leaves $(x + a) = a$, but $x = a$ or $2a = a$ and
    dividing by a leaves $2 = 1$.
    Where is the error??

Activity 20. Trip problem
    Judy is going on a trip and rain is predicted. What is the length
    or measurement of the longest umbrella she can put in her
    luggage, if the inside measurements are 15 by 20 by 10, all in

inches? What is the formula for the length if the measurements
were a by b by c?

Answers: 27 inches. Formula is $L = \sqrt{a^2 + b^2 + c^2}$ or the 3-D
          Pythagorean Theorem.

Activity 21. Activity for the day before a Federal Holiday.
          How could be the flag be designed to keep the rectangular
          shape of the Field of Stars, if there were 53 states? (This would
          be a good time to show pictures of the complete set of past
          flags. Invite an informed person to discuss the history of the
          U.S. flags.

          Answer: 8,7,8,7,8,7,8

Activity 22. How many triangles?
          This is a problem to make students think and explain how they
          solved the problem.
          Work in groups of 2 or 3.
          Question: How many triangles and what are there
                    measurements that can be formed by positive
                    integers and have a perimeter of 12?

          The two theorems used are $a < b + c$ and a > absolute value of
          $|b - c|$. Suggest they write-out these two theorems.

Suggested similar Write-out: $a < b + c$ in general means that two sides of a
                    triangle is greater than the third side.
                    $a > |b - c|$ in general means that one side
                    of a triangle is greater than or equal to the
                    absolute value of the difference of the other
                    two sides.

Suggestions for questions: Work in groups of 2 or 3.
   1. Ask for ideas as to how to start?
   2. What does perimeter mean?
   3. Write the interpretation of the two theorems?
   4. How will you use the two theorems? (This will be interesting to
       observe how they organize their work. It can be quite easy and
       systematic.)

Example: 1   10   1   This is for only one set of possibilities.
            1   9   2   All of these are invalid, except the last two!
            1   8   3
            1   7   4
            1   6   5   Valid
            1   5   5   Valid

5. Answers to the question. The number triangles is 4, and their measures are:

    3-4-5
    4-4-4
    2-6-5
    2-5-5

Students may see the need for using compasses or other means of construction to determine validity.

Activity 23. Early out

A student on Sunday afternoon was planning her busy week and didn't wish to miss school any more than necessary. Her schedule required a visit to the dentist, to the bakery, and to the market. The dentist's office is open on Tuesday, Wednesday and Friday. The bakery is closed on Wednesday and Saturday afternoons. The market is not open on Tuesday and Thursday. What afternoon should the student request early out?

Answer: Friday

Activity 24. Highest score

Who made the true statement and who had the highest score? Three students went bowling every Saturday and when asked what their averages were they made the following statements. Two confuse the situation they said that only one of the statements is true. Can you determine who made the true statement and who had the highest average? Hint: Use the indirect method.

Statements

Bob has the highest average.

Ed does not have the highest average

Dick doesn't have the highest average

Answer: Ed has the highest average and Dick made the true
           statement.

Activity 25. Summer jobs

Pete and John both had summer jobs for June, July, and August
that paid the same amount. But at the end of June Pete was
given a 10% raise and John had to take a decrease of 10% and
ironically, on August 1st Pete was given a 10% cut and John was
given a 10% increase. If in June they each made $100, then
answer the following questions.

1. What did each make in the month of August?
2. What did each make during the summer?
3. Write their summer pay if there June pay was P.

Answers: 1. 99% of P
           2. Pete made $309 John made $289
           3. Pete made $3.09P and John made $2.89P.

Activity 26. A weight problem, but a good buy!

The Athletic Department purchased six buckets of identical
tennis balls. Each bucket was from a different merchant. One
bucket contained defective balls, weight-wise. But it was a good
price so the coach decided to keep all six buckets. The coach
told the Superintendent that we can adjust to the situation, but
we need to know which bucket it is for competition.
How can he tell the under weight balls in one weighing? The
coach know that there are other ways to tell, but wanted to
challenge the team members.

Solution: Number the buckets 1 → 6 and take no balls from the #1
            bucket, 2 balls from # 2 bucket, 3 balls from #3 bucket, and so
            forth from buckets 4, 5, and 6. Put the balls you took out in a
            bag and weigh them. If the weight is correct, then you know
            the #1 bucket is defective, if the weight is 2 units short, then
            the #2 bucket is defective and so forth. Smart coach!

Activity 27. Ad Interpretation

The following are portions of actual ads. Write your interpretation of the ad in if-then form, then write the converse, inverse and contrapositive for each and state valid, invalid, or possibly valid.

Ad 1.

The NEW LIFE ad

An actor on TV played the role of a Doctor and gave a very persuasive presentation for the following. (He also knew the importance of what is called a perfect number.)

one 12 oz bottle per day plus 6 minutes of
fast walking is the formula for a
NEW LIFE.

You won't be sorry!

a. What are the perfect numbers?
b. Which words need defining? (like PURE)
c. What are some questions you would like answers

Activity 28. 1 = 0 and or ∞ = 0????

      Case 1. If 1/∞ = 0, then 1 = 0 (∞) or 0, which now reads 1 = 0

      Case 2. If 1 = 0 from case 1, then ∞(1) = 0(∞) or 0, which now
          reads ∞ = 0 or a very large number equals 0.

Activity 29. Easy Test
      Comment; Answers depend on the interpretation of the
      question. Here is a test that will look easy, but
      most students fail due to their interpretation. (A
      similar test was used by Dr. Beberman (University
      of IL in the 1950s) to prove the point.)

Questions	Answers
1. Take 4 away from 84.	8
2. Which is larger .003 or .01?	.003
3. Add 6 to 9.	69 or 96
4. What is half of 8?	0 or 3
5. What is half of XII?	VII
6. How many times does 5 go into 55?	Two
7. Which is larger 4 or 16?	4
8. Which is larger 15 or 51?	15
9. What is 64/16 equal to?	4/1
10. Which is smaller a penny or a dime?	dime

      Read answers and ask their scores, then discuss the interpretation
issue, decisions and assumptions.

Activity 30. The Binary System
      Computers use the Binary system related to the electricity as
on or off. They probably know this relationship, but can they
write the first ten counting numbers in base 2?

Suggestion: Relate the explanation to a car odometer, in Base 2, to understand the digits expressed in base 2.

As a final question suggest their write their age in base 2.

Counting numbers	Base 10	Base 2
	1	1
	2	10
	3	11
	4	100
	5	101
	6	110
	7	111
	8	1000
	9	1001
	10	1010

Activity 31. Why may a 4 legged chair wobble?

Activity 32. The Sad Romance of Miss Gon

Using the following words complete The Sad Romance of Miss Gon. Words may be used more than once.

pyramids	4	altitude	Poly	square	tangents	1	acute
base	cone		odd	relation	exponentially		parallel
2	irrational	Trape	4th-dimension		prism		inscribed
set	rings	compass	rhombus		infinity		straight
figure	irregular	right	critical point				perfect
square	face	degree	adjacent		similar		construct
empty	pi	skew	lines		point		
area	loci	limit	commute	cosine		circle	subset

THE SAD ROMANCE OF MISS GON*

Miss _____ Gon was not what you would call a beautiful woman. Southern _____ call her _____, but she had a kind _____ and was stylish to a high _____.

She met Mr. _____Zoid ____ day in the hall _____ to her father's office. Mr. _____ always stood _____ and had a good _____, although perhaps too tall and _____.

It was love at first sight for ___'____

He called on _____ at _____ times and their _____s increased ⌐_____. He always said the _____ things to her. She felt certain they were in love and at the_____ ___ of their courtship would be married, and have a _____ honeymoon.

Upon their return, she dreamed, they would _____ a house _____ to her fathers in a fashionable _____ of town. _____ would _____ to work and she would teach the _____ (math fundamentals). Her father would _____ the bank note.

To her it seemed an _____, and she reached her _____ with _____. Her life became the _____ set. She even refused to eat _____. _____ lines appeared on her face and every now and than she heaved a _____.

___ day while walking in the park she suffered an _____ pain and had to sit down at the _____ of a pine tree. Soon she began to gather _____s. Just at this _____ a messenger brought a telegram with her name _____ upon it. It announced that _____ had left her social _____ and had joined the circus to perform on the _____ and _____bars.

This shock was _____ great for _____ and she became _____. Her family blamed _____ for they always considered him in the ___ _____, and should have been in a _____.

_____ left home and departed ___ Colorado to lament the loss of her lover. But the _____ disagreed with her. She than flew to Italy and toured the country via a _____. Next she went to Egypt and while driving to the _____ one day, she lost her _____ and was never seen again.

THE SAD ROMANCE OF MISS GON
(One possible completion)

Miss Poly Gon was not what you would call a beautiful woman. Southern
tangents call her odd, but she had a kind face and was stylish to a high
degree.

She met Mr. Trape Zoid 1 day in the hall adjacent to her father's office. Mr.
Zoid always stood straight and had a good figure, although perhaps too tall
and square.

It was love at first sight for Poly

He called on Polly at irregular times and their relations increased
exponentially. He always said the right things to her. She felt certain they
were in love and at the critical point of their courtship would be married,
and have a perfect honeymoon.

Upon their return she dreamed they would construct a house similar to her
fathers in a fashionable area of town. Trape would commute to work and she
would teach their subset math fundamentals. Her father would cosine the bank
note.

To her it seemed an infinity, and she reached her limit with Trape. Her life
became the empty set. She even refused to eat pi. Skew lines appeared on her
face and every now and than she heaved a loci.

1 day while walking in the park she suffered an acute pain and had to sit down
at the base of a pine tree. Soon she began to gather cones. Just at this point
a messenger brought a telegram with her name inscribed upon it. It announced
that Trape had left her social circle and had joined the circus to perform on
the rings and parallel bars.

This shock was 2 great for Poly and she became irrational. Her family blamed
Trape for they always considered him in the 4th dimension, and should have
been in a prism.

Poly left home and departed 4 Colorado to lament the loss of her lover. But
the altitude disagreed with her. She than flew to Italy and toured the country
via a rhombus. Next she went to Egypt and while driving to the pyramids one
day, she lost her compass and was never seen again.

Add a few original lines to this story.

Activity 33. Interesting conclusions
>   Case 1.   Pick three counting numbers between 1 and 100.
>   Example: Assume 75 is picked, then proceed as follows.
>>   75 subtract 57 the palindrome of 75. Result is18.
>>   18 - 81 is -63
>>   -63 + 36 is -27
>>   -27 + 72 is + 45
>>   45 - 54 is -9

Case 2. Pick three counting numbers between 100 and 1000.
Example: Assume 579 is picked, then proceed as follows.

579 subtract 975 the palindrome of 579. Result is-396.

-396 +693 is 297

297-792 is -495

-495 + 594 is +99

Question: What will the final number of 9s be for the numbers between 1000 and 10,000. Verify your answer.

Activity 34. In Activity 9 the students found the sum of the first 100 counting numbers and also read that Gauss did the same in grade school. The challenge is to find an easy way to determine the sum of the first 100 even numbers. Let them do it on their own, in other words no suggestions from the teacher. When a student does determine the easy way, let the student explain his method and easy way to the class. Even go so far as to name the theorem after the student. Post it on the board!

Here is what the student will probably show the class.

N(The number of even numbers)		Sum
1	2	2
2	2 + 4	6
3	2+4+6	12
4	2+4+6+8	20
Eventually N	2+4+6+8+ . . . +N	N(N+1)

The student who discovered the easy way will never forget it and the theorem. He may even be called the class Gauss.

Activity 35. A Valentines Day Activity

The teacher needs to prepare in advance some polar graph paper or purchase some, and of course, it can be constructed and duplicated. (Keep the pattern and label the rays for every 15 degrees.) Calculators are advised and convenient. The objective is to graph the following.

$R = 1-\sin\theta$ for every 15 degrees (Theta is the name for the Greek letter θ)

Complete the following table and then construct the graph. (They can color it, add their thoughts, even create a card for ???.)

R	θ(in degrees)
	0
	15
	30
	45
	60
	75
	.
	.
	.
	360

If time permits challenge groups to predict and construct the graphs for:

    a. $R = 1 - \cos\theta$
    b. $R = 1 - 2\sin\theta$
    c. $R = 1 - \sin 2\theta$

CHAPTER 5

Additional Selected Activities for TGIF MATH Review

Suggestion: Try to relate these activities to practical applications common to your students' back ground or the community. This will make the activities more relevant. Let the students explain their answers and their How.

Comment: A teacher should not tell their students what to think, but teach them HOW to think.
Why or How are great words for a student to use!

1. **A proof for (-1)(-1) = +1:** (Many students are just told to memorize that the product of 2 negatives is positive, but really need to have a justification.)
   Proof:

   (-1)(-1) equals (?) This is the real question!
   (-1)(1) = -1 (1 is the identity element or number.)
   (-1) (0) = 0 Property of zero.
   (-1) [1 + (-1)] = 0 why?
   (-1)[1+(-1)] = (-1)(1) +(-1)(-1) = 0 Distributive property.
         or -1 + (-1)(-1) = 0 therefore (-1)(-1) must be = +1. QED.
         What does QED mean? (Quod Erat Demonstratum)
         See a dictionary

2. **Voting interpretations:** If 75% of the eligible voters actually voted, then 1 out of how many did not vote?

     Solution: 75% voted or 75/100 so 25% or 25/100 did not vote, which is reduced to 1/4 or 1 out of 4.

     Comment: As an easier and shorter way of writing "per cent" which represents 1/100 the symbol evolved into 0/0.

3. **What age is represented by this numeral?** The wise guy in the class said today is his $\sqrt{32}\sqrt{8}$ birthday. How old is he?
     Answer: $\sqrt{32}$ is $4\sqrt{2}$ and $\sqrt{8}$ is $2\sqrt{2}$ or the age is 16.

4. **Indirect reasoning solves the problem.**
     Some money was missing from the clubs treasury. Only 4 students had access to the funds. Only one of the 4 is telling the truth. From the following which one is it?

Student	Comment
A.	I didn't steal the MONEY.
B	I was lying.
C.	B is lying.
D.	B stole the MONEY

     Answer: B told the truth. (You may have to walk the class through the solution.)

5. **Related to lotteries: Students need lots of practice with this type of activity!** What is the probability of tossing a 3 with one toss of a die?

     a.   Ask the class to guess first and explain their guess? Answer: 1/6
     b.   Now ask their guess to toss two threes with a pair of dice? Ask for an explanation. Answer: 1/36
     c.   What is the probability of tossing a 1 or (not both) a 3 with a pair of dice?
        Let them explain their answers. Answer: 2{(5/6) (1/6) + (5/6) (1/6)} = 20/36or 5/9
     d.   What is the probability of a tossing a 3 and a 1 with a pair of dice?
        Answer: (1/6)(1/6) = 1/36 (Could be a different interpretation.)

Be sure to explain c and d and the meanings of OR and AND! There are two kinds of probability, Mathematical or Empirical. Use a pair of dice to illustrate the empirical with various cases.

6. **Understanding braking!** If a person is traveling 70 MPH in a new car and it takes 1/4 th of a second to apply the brakes, then:

   a. How far does the car travel in 1 second? Answer: 103ft nearest whole number.
   b. Now it should be easy to determine the distance in ¼ second. Answer: 26ft
   c. How far does a car traveling at 60 mph travel in 1 minute? Answer: 1 mile

7. **ACT Test questions**:

   a. What is the complement of an angle whose supplement is 120 degrees?
      Answer: 30 degrees
   b. The midpoint of a 6 inch chord in a circle is 4 inches from the center of the circle. What is the circumference of the circle in terms of pi?
      Answer: 10π
   c. Two cones have the same circular base area, but the altitude of one is 8 inches and the altitude of the smaller cone is 4 inches. If the small cone filled with ice tea cost 20 cents, then what should the large cone cost?
      Answer: Not less than 40 cents
   d. Two circular cones, the large cone has a base with a radius of 2 inches and an altitude of 4 inches. The smaller cone has a base radius of 1 inch and an altitude of 2 inches. What is the ratio of their volumes?
      Answer: VL/VS = 8/1
   e. What values for X will make this equation valid.
      3x+15 = 45
      Answer: +10 or -20
   f. What value can x not be in the following expression?
      (50X - 33)/ (2.5X - 5)
      X ≠ 2 (can't divide by 0)

Integer Problems

8. **a.** What integer meets the following three conditions?
    1. The number is less than 50.
    2. The number is a multiple of 5.
    3. The number is a perfect square.
    Answer: 25
    b. What are the missing numbers in the following sequence?
    3,9,7,13,11,?,15,?
    Answer: 17, 21

9. Which is larger, $(1/2)^2$ or 2^{-2}? Answer: They are equal.

10. What is the 2 digit prime factor of 52? Answer: 13

11. 120 is the sum of the first x counting numbers. What integer is X? Method of attack for the solution is the key. $1+2 =?$, $1+2+3 =?$ Look for the pattern!
    Answer: 15

12. a. What is the number for the value of a googol? Answer: 10^{100}.
    b. What is the largest power of 10 on your calculator? Answer: (10^{99})

Problems related to Geometry

13. What is the area of this isosceles trapezoid?

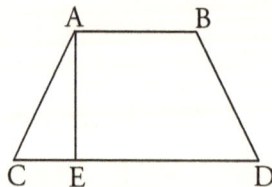

Given: AC is 5, AB is 7 Angle CEA is 90 degrees and AE is 4.

Answer: 40sq. units

14. If a regular polygon has exterior angles of 18 degrees, then how many sides does it have? Answer: 20

15. What is the ratio of the volume of the cone in figure 1 to the sum of the volumes in Figure 2?. First guess then solve for the ratio. The two cylinders are congruent.

Figure 1 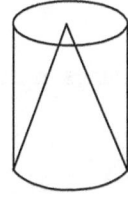 Figure 2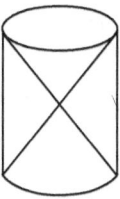

Answer: The ratio is 1/1

16. a. What is the slope of any line parallel to this line, 3x-2y=15?
        Answer: 3/2
    b. What is the x-intercept for the given line? (5,0)
    c. What is the y-intercept for the given line? (0,7.5)

Business

17. A dealer said any high school student could buy brand X shoes for this price? The $50 pair of shoes were priced at (6-8÷2)4x3-1) for a student. Was it a good buy and what is the price? Answer: yes, $22

18. **Display problem:** What is the height of this figure, if each circle has a diameter of 10?

    Answer: 25.98 units or 26 units

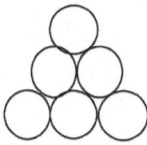

19. Jim, Bud and Greg started a summer business, where Jim put in $500, Bud $1500, and Greg put in as much as Jim and Bud together. If the business had a profit of $2400. How much should each get if the profit is shared according to the investments amounts?

    Answer: $300, $900, $1200

19. The shortest side is?

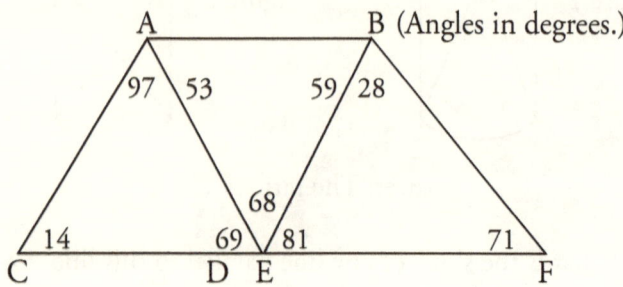

Hint: The figure is not drawn to scale! What theorem is this
          conclusion based on?

Answer: EF

A few general problems arriving at conclusions

20. The school was mailing out notices for future events and asked 32
    students to come in on Saturday to prepare the notices. They asked
    how long the chore would take? The teacher said it was known that
    4 students can do 400 notices in 1 hour. The teacher said less than
    3 hours. It actually took only 2 hours. How many notices did they
    prepare?   Answer: 6400.

21. In 2009 the local post office handled 5,000 letters per day on
    the average for the first 10 days. The rest of the month the P. O.
    handled 200,000 letters. What did the P.O. handle on the average
    per day for the rest of the month of April.   Answer: 10,000

22. A carpenter needed to cut a 10 foot 2x4 into 5 or N equal
    sections. How many cuts will he have to make for 5? For N?

    Answer: 4, N-1

23. What are the next 3 numbers in the following set, assuming they
    follow the same pattern? 1,5,10, 16, 23. _, _, _.

    Answer: 31, 40, 50

24. If the fraction A/B where A and B are real numbers, then is the answer always smaller then A? (No and justify your answer. Note: One counter example will disprove a general conclusion.)

25. What is a possible conclusion to find the sum of x odd numbers from the following?

$$1+3 = 4$$
$$1+3+5 = 9$$
$$1+3+5+7 = 16$$

Ect: Answer: x^2

Venn Diagram problem:

26. A survey reveals the following:

21 men drive Fords
25 drove General Motors cars
15 drove Chrysler products
4 were driving Ford and General Motors
3 were driving General Motors and Chrysler cars
5 were driving Ford and Chrysler cars
1 had all three company cars.

Questions.   How many drove ONLY Fords cars, only General Motors cars, only Chryslers Cars.   Answer: 12,19, 8 use Venn diagram!

27. What is a possible conclusion to find the sum of x even numbers from the following?

$$2 + 4 = 6$$
$$2+4+6 = 12$$
$$2+4+6+8 = 20$$
$$2+4+6+8+10+ \ldots x = ?:$$ Answer: The sum of the evens is $x(x+1)$.

28. a.   How many diagonals does a square have?
    b.   How many diagonals does a regular pentagon have?

    c. How many diagonals does a regular hexagon have?

    d. How many diagonals does a regular heptagon have?
Do you see a pattern and predict how many diagonals a regular octagon has?   Answer: D = N(N-3)/2 where N is greater than 3.

29. **Prediction or induction problem.**

    5 x 5 =25
    15 x 15 = 225
    25 x 25 = 625
    35 x 35 = ?   Predict and then ck your answer using calculator.
    45 x 45 = ?   Answer: 2025 or (4x5)25
    Write a conclusion.

ACT Test problem:

30. Draw a 30-60 degree right triangle and indicate the altitude from the vertex of the right angle.

    a. Indicate the size of all 6 angles.

    b. Map the three similar triangles.

    c. If the hypotenuse is 10 units then calculate the lengths of all the segments.

Decision Making Review

As mentioned before the teaching of Mathematics has two major objectives:

1. Teaching the mathematics needed for whatever the professional background calls for. This means being a student is a life time obligation.

2. The skill of Decision Making for a better life. This was stated at the entrance to Plato's Academy as: **LET NO MAN ENTER HERE IGNORANT OF GEOMETRY**.
To meet the needs of modern day requirements this has been changed to: **LET NO STUDENT EXIT HERE IGNORANT OF GEOMETRY.**

The problem is that not many Geometry teachers teach their course with this objective.

31. Ask the student to list what they think decisions are based on. A few items may be:

> What they see. (TV)
> What they read.
> What somebody told them.(Is the person an authority?)
> National Polls (What information should be stated? See below.)
> Methods of logic, direct and indirect reasoning.
> Bias
> Tradition
> Etc and keep a list for future reference.

An example of the use and problem with polls recently in Time magazine. January 28, 2013 issue.
The following was stated with regard to the issue of Gun Control:

> Would you favor a background check?
> 92% at the store 87% at gun show 75% private seller
> Would you favorite the following for gun safety?
> 69% gun registration 58% ban on clips
> 56% ban on assault guns 52% ban on Ammo sells
> Would you favor or oppose armed guards in schools?
> 54% favor 45% oppose
> Do you own a gun?
> 49% Yes 49% no
> Who is to blame for gun violence in the USA?
> 37% parents 37% pop culture 23% availability of guns

This is only a part of the survey, but what is very important in evaluating the results of a survey was stated in very fine prints and hard to read is the following.
The survey was listed as international, conducted by phone on Jan. 14& 15 (Sunday and Monday) of 814 adult Americans selected at random. Would you draw a general conclusion base on 814 adults?

What would you like to know about these 814 adults?

1. Were they Gun owners?
2. Where they live?
3. Age
4. Profession or career.
5. Education
6. What is their favorite news show?

Comment: Do cars kill? (You must qualify to drive.)

Class should discuss the responses to the above!

32. If you are a good teacher, then you like students (Assume this to be true) Which of the following are valid? (Remember: Statements are true or false and conclusions are valid or invalid.)

   a. John is a good teacher, then he likes students.
   b. Peter likes students, then he is a good teacher.
   c. Huck doesn't like students, the he is not a good teacher.
   d. Midge isn't a good teacher, then she doesn't like students.

   Answers: V, I, V. I (In order a-d)
   Teacher: Use A→ B diagram to show the relationships for the converse, inverse, contrapositive.

33. From the following statement write two conclusion that are valid and two that are invalid.
   The State Good Driver Association indicated that very few drivers were killed at speeds over 100 mph.
   Letters to the editor stated the following conclusion.

      a. (your valid conlusion.)
      b. (your valid conclusion)
      c. (your invalid conclusion.)
      d. (your invalid conclusion)

34. What is inductive reasoning? The following will illustrate why you need to be careful when arriving at a conclusion by inductive reasoning. (Problem credited to Leo Moser)

a. Draw a circle on your paper with at least a 2 inch diameter.
b. Select 2 points, say A and B, on the circumference and draw segment AB.
    Circle is divided into how many regions. Conclusion: 2 points, then 2 regions.
c. Pick another point, say C, on the circumference and draw the chord.
    How many regions now? Conclusion: 3 points and __ regions.
d   Pick another point, say D, on the circumference and draw the chord.
    How many regions now? Predict: Conclusion: 4 points and __ regions by counting.
e. Pick another point, say E, on the circumference and draw the chord.
    How many regions now? Predict? Conclusion: 5 points and __ regions by counting

Was you prediction correct? What does this tell you with regard to inductive reasoning?

35. Indirect Reasoning How did A know he had a black hat on?
    Three students had a GPA of 4.0 and the class asked the teacher which one he thought is the most intelligent? The teacher set up the following test. He instructed each of the three to stand in a corner of the room. He then **blind folded** each of them and carefully explained he would put a **black or white** hat on each of them. When the blind folds are removed they are to put their right arm above their head if they see a black hat and they are to take their hand down when they know the color of their hat. He then put a black hat on each and took off the blind folds. They each put their hand above their head. After a short while one student took his hand down and said I have a black hat on! How did he or she know?

Suggestion: Have three black hats and act this out in class for complete understanding! This can take the whole period.

BIBLIOGRAPHY

Suggestions for further reading

Abbott, Edwin A.
FLATLAND A ROMANCE OF MANY DIMENSIONS
Princeton University Press

Banks, Robert B.
SLICING PIZZAS, RACING TURTLES, AND FURTHER ADVANTURES IN APPLIED MATHEMATICS
Princeton University Press

Beckmann, P.
HISTORY OF PI
Golen Press

Bell, E. T.
MEN OF MATHEMATICS
Simon & Schuster

Byrkit, D.
"TAXICAB GEOMETRY."
MATHEMATICS TEACHER, May 1971, Pages 418-422

Cajori, Florian
HISTORY OF ELEMENTARY MATHEMATICS
The Macmillan Company

Davis, J.J.
   BIBLICAL NUMEROLOGY
   Baker Book House (PI value is stated in the Bible, erroneously, I Kings
   7:23)

Davis, P. and Hersh, R.
   THE MATHEMATICAL EXPERIENCE
   Houghton Mifflin

   DESCARTES DREAM
   Harcourt Brace Javanovich

Devlin, K
   Mathematics-the new golden age
   Columbia University Press

Dudley, Underwood
   NUNEROLOGY or, What Pythagoras Wrought
   Mathematical Association of America

   MATHEMATICAL CRANKS
   Mathematical Association of America

Fadiman, Clifton
   THE MATHEMATICAL MAGPIE
   (Mobius Strip—"Paul Bunyan vs. The Conveyor Belt)"
   Simon and Schuster

Fawcett, Harold
   NATURE OF PROOF
   13th Yearbook of NCTM

Florman, S. C.
   ENGINEERING AND THE LIBERAL ARTS
   McGraw-Hill Company
   (A guide to History, Literature, Philosophy, Art,
   Science, and Music)

Gardner, M
MATHEMATICAL CARNIVAL
Alfred A. Knopf

MATHEMATICAL CIRCUS
Vintage Books
Division of Random House

Gazale, Midhat
NUMBER: From Ahmes to Cantor
Princeton University Press

Gordon, Sheldon and Florence, Editors
STATISTICS FOR THE TWENTY-FIRST CENTURY
Mathematical Association of America, 1992

Huff, Darrell
HOW TO LIE WITH STATISTICS
Norton Co.

Kenny
"Hemholtz And The Nature Of Geometric Axioms"
Mathematics Teacher, Vol. 50, Feb. 1957

Klein H. A.
THE WORLD OF MEASUREMENTS
Simon and Schuster

Kline, M.
MATHEMATICAL THOUGHT FORM ANCIENT TO MODERN TIMES
Oxford University Press

Lieber, L.
MITS, WITS, AND LOGIC
Institute Press, New York, 1954

THE EDUCATION OF T. C. MITS
W. W. Norton & Co., 1954

Loomis, E.
THE PYTHAGOREAN PROPOSITION.
NCTM publication
Comment: (Which former President of the U.S. is credited with a proof?)

Nolan, Deborah, Editor
WOMEN IN MATHEMTICS: SCALING THE HEIGHTS
Mathematical Association of America

Northrop, E. P.
RIDDLES IN MATHEMATICS (A Book of Paradoxes)
D. Van Nostrand Company

Packel, Edward
THE MATHAMATICS OF GAMES AND GAMBLING
Mathematical Association of America

Paulos, J.
I THINK, THEREFORE I LAUGH
Vintage Books
Division of Random House

Peterson, I.
THE MATHEMATICAL TOURIST
W. H. Freeman and Company

Poe, Edgar Allen
THE GOLD BUG (A Mystery involving mathematical reasoning.)

Polya, G.
MATHEMATICAL DISCOVERY: Vol. 2 (Chapter 14: The art of teaching mathematics.)
John Wiley & Sons

Postman, N.
TECHNOPOLY
Alfred A. Knopf

Reid, Constance
A LONG WAY FROM EUCLID
Thomas Y. Crowell Co.

Reeve, W. D.
THE TEACHING OF GEOMETRY
5th Yearbook NCTM

Stevenson, R. L.
TREASURE ISLAND (chapter 31)
(Locus problem-location of the treasure.)

Weber, R.
A RANDOM WALK IN SCIENCE
Crane, Russak & Co. Inc.
"Life on Earth.(by a Martian")
(Fascinating little story (p. 124) with a surprise ending.

Video or film
DONALD DUCK IN MATHMAGIC LAND
Disney

An interesting critical thinking test.
Critical Thinking Test, Level X
R. Ennis and J. Millman
(Very interesting and a different type of test based on a space travel theme. I have given this to several hundred high school and college students on a pre/post test situation and to my surprise the average group gained the most.)
Available at:
Foundation for Critical Thinking 1-800-833-3645 or 1-800-458-4849
www.criticalthinking.org

JIM ELANDER

Web sites

www.//history.mcs.st www.MAA.org
www.archives.math.utk.edu/societies.html
www.nsf.gov/ www.AMS.org/ www.forum.swarthmore.edu/ncsm
http://Turnbull.mcs.st www.history.mcs.st-andrews

Also use computer search for "Math History" or "Math Archives"

QUOTES

MATHEMATICS IS THE GATE AND THE KEY TO ALL SCIENCES.
HE WHO IS IGNORANT OF IT CANNOT KNOW THE THINGS
OF THIS WORLD.

Roger Bacon

YOUNG PEOPLE WHO HAVE ACQUIRED THE ABILITY TO
ANALYZE PROBLEMS, GATHER INFORMATION, PUT THE
PIECES TOGETHER TO FORM TENTATIVE SOLUTIONS WILL
ALWAYS BE IN DEMAND.

J. G. Maisonrouge
Board Chairman
IBM World Trade Corp.

That they (all citizens) might excel in public discussions on philosophic
or scientific questions, they must be educated(rhetoric, philosophy,
mathematics, and astronomy).

The Athenian Sophist School
Curriculum(480 B.C.E.)
F. Cajorie

CONSCIOUSLY MATHEMATICS HAS BEEN A HUMAN ACTIVITY FOR THOUSANDS OF YEARS. TO SOME SMALL EXTENT, EVERYBODY IS A MATHEMATICIAN AND DOES MATHEMATICS.

Phillip Davis & Rueben Hersh
THE MATHEMATICAL EXPERIENCE

GOD gave us the integers (whole numbers) and all the rest is the work of man.

L. Kronecker

Mathematics is not a spectator sport!

Anonymous

Neglect of mathematics works injury to all knowledge.

Roger Bacon

Number rules the universe.

The Pythagoreans

Mathematics—the unshaken Foundation of Sciences, and the plentiful Fountain of Advantage to human affairs.

Issac Barrow

Understanding evolves from work, appreciation from applications.

Unknown

The theory of probability entered mathematics through gambling.

P. Davis & R. Hersh
THE MATHEMATICAL EXPERIENCE

In truth, all of life in one way or another is concerned with the study of probability.

> H. Gross & F. Miller
> *MATHEMATICS—A Chronicle of Human Endeavor*

In short, the house plays the percentages, while the player relies on luck . . .

> H. Gross & F. Miller
> *MATHEMATICS—A CHRONICLE OF HUMAN ENDEAVOR*

All numbers in the form of 4n+1 are the sum of 2 squares.

> Fermat

A mathematician, like everyone else, lives in the real world. But the objects with which he works do not. They live in that other place—the mathematical world. Something else lives here also. It is called TRUTH.

> Jerry P. King
> *THE ART OF MATHEMATICS*

You cannot fake. In mathematics, no one can be fooled. You can either prove . . . or you cannot.

> Jerry P. King
> *THE ART OF MATHEMATICS*

Many of the laws of the sciences are stated in the language of variation.

> Unknown

Mathematics is like a mighty tree with number (counting numbers) for its roots. Arithmetic grows on numbers, algebra on arithmetic, geometry on arithmetic and algebra, analytic geometry on arithmetic, algebra, and geometry. Calculus builds on all four. It is a tree that grows in time, fertilized by the minds of mathematicians and the applied needs of society.

Unknown

Attributing teaching and learning failure to something called "math anxiety" serves no purpose except to provide a built-in excuse for inadequate performance on both sides.

Jerry P. King
THE ART OF MATHEMATICS

Statistical thinking will one day be as necessary for efficient citizenship as the ability to read and write.

H. G. Wells

Pythagoras, the teacher, paid his student three oboli (a coin) for each lesson he attended and noticed that as the weeks passed the boy's initial reluctance to learn was transformed into enthusiasm for knowledge. To test his pupil Pythagoras pretended that he could no longer afford to pay the student and that the lessons would have to stop, at which point the boy offered to pay for his education . . . \

Simon Sing
FERMAT'S ENIGMA

TO MEASURE IS TO KNOW

Johann Kepler

Hipparchus of Nicaea, (180-125 B.C.E.) compiled the first trigonometric table.

Boyer, C.B.
A HISTORY OF MATHEMATICS

The advance and the perfecting of mathematics are closely joined to the prosperity of a nation.

Napoleon

The heart of the mathematical experience is, of course, mathematics itself.

Davis, P and Hersh, R
THE MATHEMATICAL EXPERIENCE

Mathematics through the power of computers pervades almost every aspect of our lives . . .

David L. Goines

To Think is to Know

Unknown

Let no man ignorant of Geometry enter here.

Plato

Let no person ignorant of Geometry exit here.

J. Elander

Students of mathematics . . . the first time something new is studied seem they hopelessly confused . . . Then, upon returning (to the concept) after a rest, . . . everything has fallen into place.

E. T. Bell
MEN OF MATHEMATICS

Descartes . . . the essence of plane analytic geometry lies in the matching of ordered pairs of real numbers with the points of a plane.

> Edna E. Kramer
> *THE NATURE AND GROWTH*
> *OF MODERN MATHEMATICS*

Many of the laws of the sciences are stated in the language of variation.

> Unknown

Thinkers recognize when two variables are related, but it is Mathematics that connect them numerically.

> Unknown

Mathematics is like a mighty tree with number (counting numbers) for its roots. Arithmetic grows on numbers, algebra on arithmetic, geometry on arithmetic and algebra, analytic geometry on arithmetic, algebra, and geometry. Calculus builds on all four. It is a tree that grows in time, fertilized by the minds of mathematicians and the applied needs of society.

> Unknown

There cannot be a language (mathematics) more universal . . . and more worthy to express the invariable relations of the natural things.

> Joseph Fourier

Analytic Geometry . . . constitutes the greatest single step ever made in the progress of the exact sciences.

> John Stuart Mill

The essence of plane analytic geometry lies in the matching of ordered pairs of real numbers with the points of a plane.

Edna E. Kramer

I THINK THEREFORE I AM.

Rene Descartes

The Great Architect of the Universe now begins to appear as a pure mathematician.

J.H.Jeans
The Mysterious Universe

The definition of a good mathematical problem is the mathematics it generates rather than the problem itself.

Andrew Wiles

We learn the new in the light of the old.

Anonymous

The important thing is to not stop questioning . . .

Albert Einstein

Relationships between different subjects (even branches of mathematics) are creatively important in mathematics.

Simon Singh
FERMAT'S ENIGMA

Mathematics consists of islands of knowledge in a sea of ignorance.

Simon Singh
FERMAT'S ENIGMA

A mathematician, like everyone else, lives in the real world. But the objects with which he works do not. They live in that other place—the mathematical world. Something else lives here also. It is called TRUTH.

Jerry P. King
THE ART OF MATHEMATICS

Statistics makes possible new perceptions and realities by making visible large-scale patterns.

Neil Postman
*Technopoly—The Surrender
of Culture to Technology*

Statements are true or false. Conclusions are valid or invalid

Unknown

Just as statistics has spawned a huge testing industry, it has done the same for the polling of "public opinion."

Neil Postman
*Technopoly—The Surrender
of Culture to Technology*

It is not how much you cover, but how much you uncover.

H, Fawcett

The connection between the improvement of human conditions and the happiness of the human race is Science. (The Queen of the sciences is MATHEMATICS.)

Neil Post*man*
Technopoly-The Surrender
of Culture to Technology

The proof of the pie is in the eating.

Unknown

There is no royal road to Mathematics

Menaechmus
(to Alexander the Great)

Mathematics is the science of making necessary conclusions.

B. Peirce

It is easier to square the circle then get around a mathematician

A. DE Morgan

Mathematics is about anything as long as it is a subject that exhibits the pattern of assumption—deduct-conclusion.

P. Davis and R. Hersh
The Mathematical Experience

www.ingramcontent.com/pod-product-compliance
Lightning Source LLC
Chambersburg PA
CBHW022108170526
45157CB00004B/1540